Industr

INDUSTRY4.0 *for* PROCESS SAFETY HANDBOOK

JIM PETRUSICH
HANS VOLKMAR SCHWARZ

ISBN: 978-1979021166

Contents

Preface

This book is intended for professionals working in hazardous process industries such as oil & gas, chemical, and specialty materials. The purpose of this handbook is to provide a guide for setting up a collaborative platform which will reduce process safety risk by leveraging existing data (sensors, processes, etc.) with analytics, the foundation of Industry 4.0.

To be successful organizations must cultivate a process safety culture. Many companies confuse this with a focus on personal safety, but although related the topics are as different as football and cooking.

Those at all levels in an organization who are interested in process safety can leverage Industry 4.0, and use this book to help promote the culture. The book and chapters are intentionally short and to the point. A chapter could be used as a topic to start a company meeting, or the book could be used to outline a corporate initiative.

Introduction

In February 2015 a large explosion occurred at a refinery in the Los Angeles area. Although this was a major event it was also considered a near miss, and according to Vanessa Allen Sutherland, Chair of the Chemical Safety Board (CSB), *"had the potential to be catastrophic."* A large piece of debris from the explosion narrowly missed hitting a modified hydrofluoric acid tank (highly toxic). The CSB cited multiple gaps in the refinery's process safety management system including *"the refinery relied on safeguards that could not be verified."* Workers at the refinery were running *"blind"* leading up to the event. (Penn)

In August 2015 an explosion at a chemical storage station near Beijing, China killed 173 and injured 797. The first of ten explosions registered as a 2.3 magnitude earthquake…the second explosion registered 2.9. Poor decision guidance may have caused the next eight explosions three days later. Firefighters used water to combat the initial fire which can cause calcium carbide to release acetylene (a highly volatile gas), and may have detonated the ammonium nitrate stored on site. Of the 173 killed, 106 were first responders (95 were firefighters and 11 were police officers). (Iyengar)

In September 2017 a chemical plant near Houston experienced major explosions from stored organic peroxide which lost refrigeration. The plant had lost power in the flooding from hurricane Harvey. This handbook was published a month later.

Chemical disasters continue to happen. If we want to make these incidents a thing of the past we need to challenge our current approach. The industry operates in high risk environments and companies design extensive processes to mitigate significant safety threats. With the advent of Industry 4.0 many of the current practices need revisiting. New approaches will leverage data far better and build on the foundational work from the past. The current systems in use developed diverse data silos *(historians, LIMS, etc.)* with challenges to access and aggregate an overall understanding. With the rapid expansion of powerful instruments, data collection and automated analysis, the opportunity to develop a unified approach to monitor process safety is available.

In most cases, major incidents are preceded by signals which, if recognized, would enable disaster to be avoided. After Three Mile Island, Charles Perrow wrote that these events are *"unexpected, incomprehensible, uncontrollable and unavoidable"* in his book "Normal Accidents." Perrow's statement may have been true when his book was published in 1984, as this was before sensors, large scale data storage, and analytic monitoring had been revolutionized. Today this is no longer the case. Industry 4.0 enables us to move beyond this viewpoint.

In 1984 the cost of a gigabyte of storage was about $100,000 USD; today it is under 10 cents. (Komorowski) We need to be careful in how we handle this windfall, as data and information are not the same. This book will focus on monitoring and analyzing data to create information for decision guidance.

A company focused on becoming a High Reliability Organization (HRO) now has the ability to develop an analytic monitoring platform that integrates all instruments and data sources with decision guidance to avoid catastrophes. This needs to be incorporated into company culture as we are now developing the Digital High Reliability Organization (DHRO).

CHAPTER 1

Process Safety: ROI

From a financial perspective, safety is one of the top Key Performance Indicators (KPIs) for any company operating in hazardous environments. Increasing shareholder value depends on how operations perform both from an operating efficiency and process safety perspective. Plant acquisitions and divestitures are common. Having a strong statistical approach to understanding and incorporating the safety risks of newly acquired operations is essential.

There are several areas of process safety risk: environmental, regulatory, human safety, plant assets, corporate image, and litigation. A disaster will frequently affect several or all of these categories. An explosion in one plant may cause loss of life, the destruction of major company assets, an environmental impact, negative corporate image, regulatory compliance issues, and law suits from those affected by the event.

After an incident, the plant is often not operable for extended periods of time. That means lost business. Typically that is a much bigger loss than the direct mechanical damage in the plant. An explosion in a plant could result in $50 million of direct damage to repair, but $300 million in lost contribution margin from not being able to produce for the time until the damage is repaired. The impact could be even larger if customers who were forced to switch suppliers do not come back.

For most of these variables the return on investment is based on measuring the incidents you avoid. This can be difficult for a company that has never had a major disaster. How do you measure something which has never happened? However you can measure all loss of containment incidents that have happened, and there are effective ways to assess the risk of a major catastrophe.

Corporate Dartboard

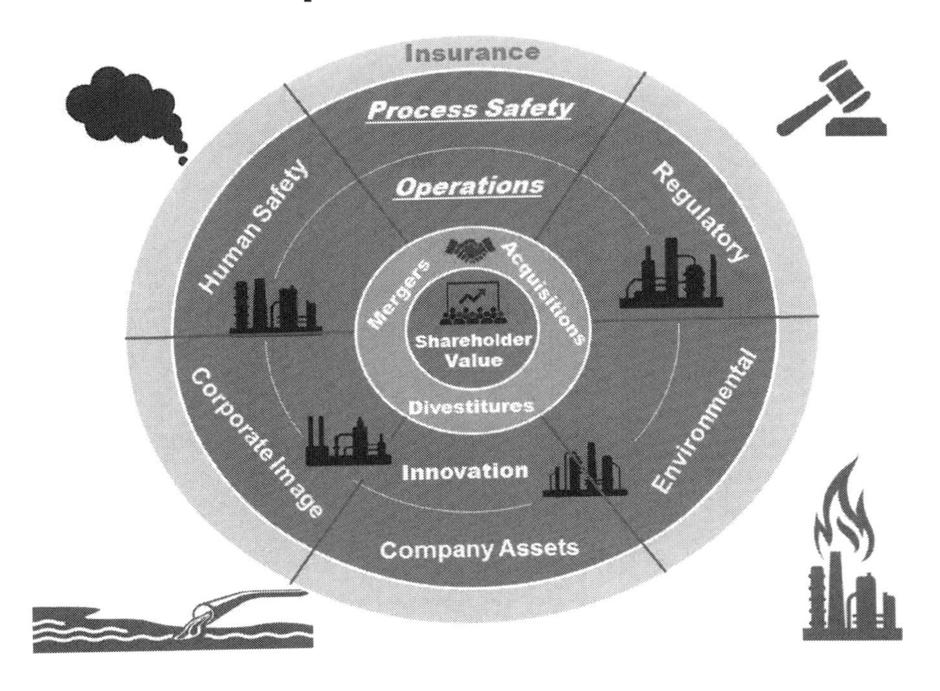

If increasing shareholder value is the target on a corporate dartboard, process safety incidents are what happens when companies are not mindful. Without deliberate effort organizations miss the dartboard altogether and disaster occurs. The ramifications can be tremendous. Having successful mergers, acquisitions, and divestitures is essential. Driving innovative approaches to operational efficiency and managing process safety risk is how this happens.

Many insurance companies are now providing significant discounts to companies who are achieving process safety through Industry 4.0 initiatives. These companies show tangible benefits from automating the Industrial Internet of People *(IIoP),* and Things *(IIoT).*

Effective operations and plant safety are intertwined. A focus on reducing process safety risk usually has a positive impact on overall plant efficiency *(less variation = less risk)*.

Fewer unplanned shut downs means fewer startups, which is the most common time when major incidents occur. Improving Overall Equipment Efficiency *(OEE)* will also increase process safety.

After deploying a holistic near real-time process safety system, Scottish Power saw a 50% drop in their process safety incidents in less than one year. At the same time they experienced a significant drop in capex and opex while increasing plant optimization. The following results were achieved over a three year period:

- 22% increase in plant availability
- 29% decrease in operations and maintenance costs
- 50% reduction in unplanned outages
- 10% decrease in insurance costs (Martin Sedgewick)

CHAPTER 2

Why Industry 4.0 For Process Safety?

How are the signs which could prevent disaster missed with plant technology and modern Advanced Process Control (APC) in place? Why can operations personnel still be blind to the dangers? Hard systems, such as pressure relief valves and plant trips, have been put in place to prevent these issues. So why do Process Safety Incidents (PSIs) continue to happen?

1. A Process Safety Incident is five times more likely to happen during startup than any other time. During startup most APC systems are turned off and plant processes are run in manual mode. Usually much of the alarming is either turned off or ignored, as these signals are designed to monitor variation once the process is in steady state.

2. Sometimes the optimal target is achieved when running close to the plant trip point *(because of output).* This tends to create more shut downs, and more shut downs naturally create more startups.

3. Although a great deal of information is captured about the process, much of this is in various data silos, and too little is aggregated and analyzed to monitor overall process safety risks.

Some people may point out that human error is frequently the cause of failures which have catastrophic consequences. We should expect that humans will occasionally make poor judgements. Rather than assume these errors are inevitable (not very useful), the focus should be to move upstream on the data chain to provide early warning and actionable guidance to prevent issues. This is entirely possible.

Plants have accumulated enormous amounts of data over the last 10 to 20 years in their Distributed Control Systems (regarding the plant operation) and in their maintenance systems (regarding what gets repaired how often and for what reason). These data combined can tell you which parts of the plant are vulnerable to slip into a lower stability zone and are prone to more problems.

Data may also tell which parameter set is optimal to avoid issues, and may help predict the next problem, so it can be avoided. Advanced analytics makes use of all these data even if they are only rarely used.

Some plants are now being built in a virtual environment. Prior to start up operators can "walk the line" either electronically or physically with guidance to identify the right valve position, temperature setting, flow rate, etc. These augmented reality methods help operators with decision guidance while reducing errors and the risk of confusion.

CHAPTER 3

Process Safety vs. Personal Safety

The common Injury Triangle was developed in 1931 by Herbert Heinrich while working at Travelers Insurance. His research showed a consistent direct pyramid-like relationship for the number and severity of injuries. Multiple studies have been conducted over the decades since and all confirm the relationship of minor to major accident rates. From Heinrich's research he developed the theory of Behavior-based Safety which focuses on encouraging safe actions to prevent safety incidents. (Heinrich)

Many companies operating hazardous processes inferred from this relationship that focus on safe behavior will prevent major catastrophes. This is not correct.

Personal safety can be monitored by the things you see: Are employees wearing their protective gear? Do they use the hand rails? Do others intervene when they see unsafe acts?

Process safety issues frequently are in areas you cannot see: What are the critical temperatures and pressures? Are the valve positions correct? Where are our tank levels? How are subject matter experts and operators collaborating? ...What about during shift changes?

We will review five past disasters in Chapter 8, all of which with companies that focused on safety. Most of the attention was focused on minor injuries or lost-time injuries, and they had impressive records for these metrics. Two of the companies we review were presenting safety awards on site the day the catastrophe occurred. *(Texas City Refinery, Deep Water Horizon)*

The Two Triangle model addresses the difference between process safety and personal safety: Personal safety practices can prevent isolated incidents. Major disasters occur when (frequently unseen) process variables go out of control. Complex relationships often cause a domino effect, building to a catastrophe despite operators having benchmark personal safety practices.

There certainly is some overlapping benefit to an overall culture of safe behavior, but monitoring critical process parameters is the essential component for reducing the risk of a major disaster. Companies, who thought focusing on the injury triangle would prevent major incidents, have been proven wrong over time. Focusing on personal safety is not wrong; it is simply not enough.

Lost time injury is a good metric to measure personal safety. Process safety should be measured by different metrics: such as the number of unplanned shutdowns, emergency relief systems activated, or loss of containment incidents.

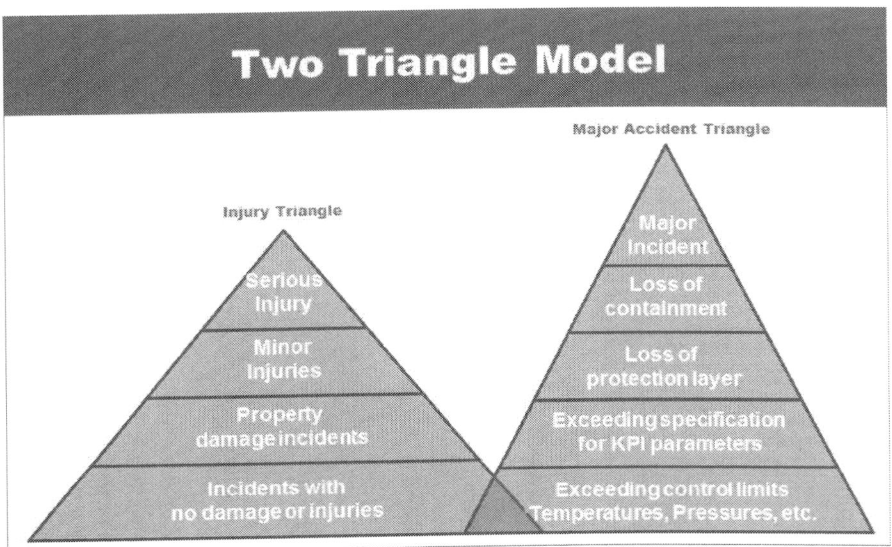

Physically looking for process safety hazards is important but there is risk in relying too much on an inspection scheme for asset integrity. Corrosion and equipment fatigue could occur in less expected places.

The 2015 refinery explosion in the Los Angeles area occurred when undetected corrosion in a closed valve allowed all of the catalyst to leak. The engineering team was unaware of the condition, and executed start up procedures which initiated a series of reactions that led to the explosion. (CSB, video)

Process safety relies heavily on data from sensors and instruments. Having a system for assuring the precision and accuracy of this data is critical to process understanding and making the right decisions.

CHAPTER 4

Playing With Matches

"The first principle is that you must not fool yourself ...and you are the easiest person to fool."
-Richard Feynman, Nobel Prize Winner

Experts are critical to breaking down and understanding process safety. They have the knowledge to frame the problems which exist in complex chemical manufacturing processes. They also know the process safety questions to ask and where the important data resides.

Experts should be used to design the measurement system, and allow the data analytics to determine the results. This may sound obvious, but the process safety industry uses tools like the Risk Matrix (Chapter 7) which relies on expert judgement. There are several flaws with this methodology which will be described later. In many cases even experts jump to conclusions without fully interrogating the data.

Why do experts keep telling this story wrong?

The following story was told in a Life Magazine article by evolutionary biologist, Julian Huxley in 1952. Carl Sagan retold the story in his popular television series *"Cosmos,"* and also in his book by the same name in the 1980s. Richard Dawkins retold the story in 2009 with his book *"Greatest Show on Earth."*

According to Japanese legend, fisherman consistently threw back certain crabs that reminded them of the samurai warriors lost in one of the most famous battles in Japanese history. This strange crab has a carapace shell with the appearance of a samurai warrior and this distinctive look spared them from being eaten by humans.

Scientists have used this story to show how evolution works, as natural selection migrates to the attributes which best leads to survival. This case has been referred to as *"artificial selection"* rather than *"natural selection"* since the crabs might not have been the strongest or fittest, but were ignored because of human pattern recognition.

The Emperor Antoku's 1,000 ship fleet of Heike Samurai battled the Genji Samurai with their 3,000 ship armada.

When the battle ensued, Emperor Antoku was only 6 years old. He was considered wise for his age, but was heavily guided by his grandmother. The Heike Samurai fought to the death with swords and daggers, but none survived the day.

Preferring death to capture, Antoku's grandmother wrapped her robe around him, and led him into the sea telling him "there is another kingdom beneath the waves." The few women on the emperor's ships who were allowed to live became part of the Genji Samurai's Miyamoto Imperial Court. The descendants of these women continue to celebrate the anniversary this famous battle every April 24th.

According to legend, for allowing their clan to be defeated, the ghosts of the Heike Samurai were destined to roam the bottom of the ocean in the form of crabs. Japanese fisherman would catch these crabs and throw back the ones which appeared to have the face of a Samurai warrior on the shell. These crabs are known in Japan as the Heike-gani.

The naval battle of Dan-no-ura in 1185 ended the era of the aristocrats and began the era of the warrior class Shoguns, who would rule Japan for nearly seven hundred years.

When well-known evolutionary biologist, Julian Huxley, retold the story for Life magazine he stated this was a case of natural selection through deliberate human sorting.

> *"…the resemblance …to an angry Japanese warrior is far too specific and far too detailed to be accidental."*

Carl Sagan later wrote:

"These crabs became victims of natural selection, as the ones that resembled the look of a Samurai warrior face, would be spared. What started off being random, became the designed look, and by the process of natural selection, or really 'artificial selection' created a common appearance."

Although this is an amazing story, these top scientists were completely wrong.

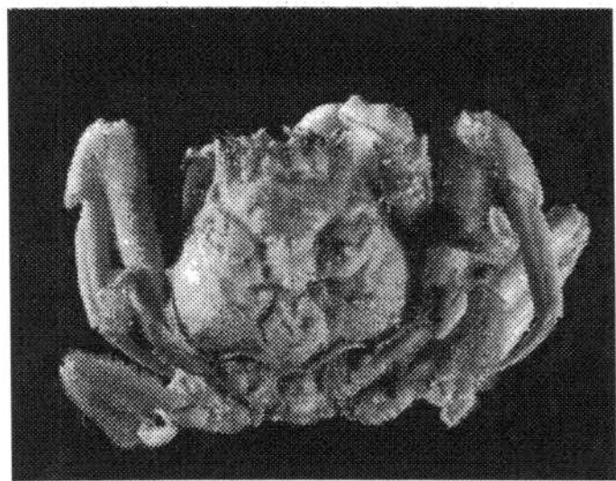

Joel Martin is a marine biologist who specializes in crustaceans at the Natural History Museum of Los Angeles. He shows that the Heike crab *(Dorippe Japonica)* has the look of a scowling samurai face, but so do others in the Dorippidae crab family located all over the world. In fact the crab pictured above the story is Paradorippe Granulata, which lives in waters of the Pacific Northwest (off North America).

The Heike Crab is pictured below.

Japanese fishermen do not eat any of these crabs. They throw all of them back whether they resemble a samurai face or not. Finally, Martin points out that crab fossils which predate man also have the appearance resembling a samurai warrior face. (Martin, The Samurai Crab)

Scientists like Sagan and Huxley see thousands of similar relationships throughout their careers. The temptation is to jump to a conclusion before thoroughly evaluating all the available data. The Heike Crab history is just folklore, and not a story of natural (or artificial) selection.

Plant folklore exists in most process manufacturing sites. Something bad may have happened in the plant at one point in time, and some parameter was unusually high or low. That relationship was interpreted as causal and a new process was put in place based on this false assumption.

An analytic monitoring system equips subject matter experts with a superior decision-making tool. They know the important variables, the context, the relationship among the variables, and how to ensure data integrity. They know what should be tested, and the possible decisions to be made based on the results.

Many of the top companies in the industry are focused on becoming High Reliability Organizations (HRO). These companies challenge current assumptions and continually look to catch any new threats as they emerge.

"If you wish to see the truth, then hold no opinions."

-Zen Proverb

CHAPTER 5

Digital High Reliability Organizations

HRO: An organization which can avoid catastrophes in an environment where they are expected.

HROs deliberately create a culture of responsibility to provide far more attention to potential safety threats. They continually interrogate the definition of what is hazardous. Every stakeholder must be responsible to understand the current situation and the impact of any potential actions. An Industry 4.0 platform provides the structure to enable a HRO, but requires the commitment to a mindful safety culture.

One way to view this is by looking at the four stages of learning. When we first face a problem, we are unaware, until suddenly recognizing the issue. This is when we move from *"unconscious incompetence"* (we don't know what we don't know) to "conscious incompetence" (we recognize the problem).

Think of learning to tie your shoes: young children at some point learn to dress themselves except for one part. They realize that someone needs to tie their shoes. They have moved to conscious incompetence. Then someone shows them how to tie their shoes and if they concentrate they can learn to do this themselves. They have moved to conscious competence. After practice tying shoes becomes intuitive and they no longer need to think about it, they move to unconscious competence.

This works well for simple tasks like tying shoes, but is disastrous for complex problems such as process safety. HROs focus on staying in the consciously competent stage. In other words, *"we know what we are doing and we know why we are doing it."*

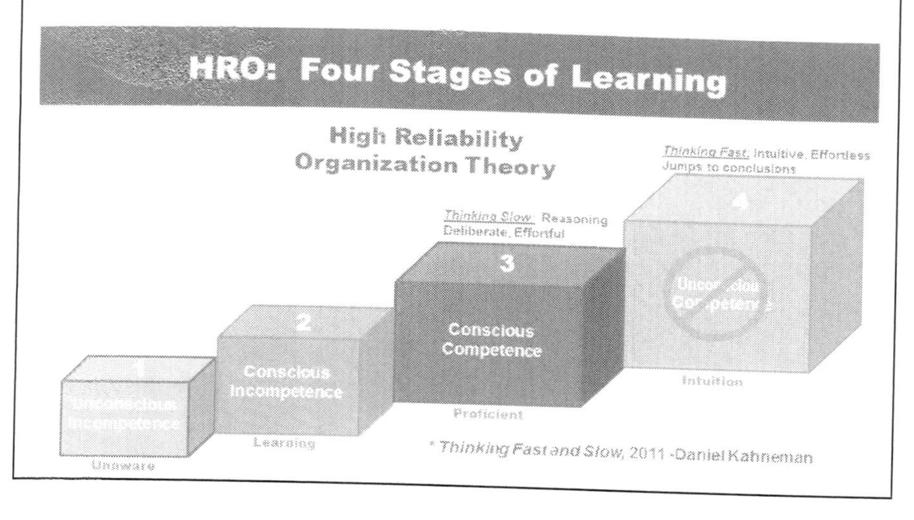

If you were to summarize how HROs function in a word, they operate with "mindfulness." The organization is structured with training, monitoring, and auditing to support a mindful approach. *"Mindfulness preserves the capability to see the significant meaning of weak signals and give strong responses to weak signals."* (Karl Weick) This goes beyond just the individual mindset. The culture and structure of the organization must create an environment where mindfulness is an integral part of the process.

HRO theory is supported by many studies. Some were conducted by Daniel Kahneman who won the 2002 Nobel Prize for his research in Behavioral Economics. Along with his research partner, Amos Tversky, Kahneman found that many people made bad intuitive decisions on "gut impulse." If provided with better data people made more thoughtful deliberate decisions which were more rational. Interestingly, more data did not necessarily lead to better decisions. Appropriately filtered data shown in an appropriate context led to better decisions.

Mindful HROs should consider leveraging all the available data to drive information and knowledge. This means driving a Digital HRO culture by guiding decisions with an analyzed data platform. By tracking actions and how the process responds to different situations, process safety leaders can drive continuous improvement.

Operators are guided with handheld devices when they "walk the line" which identify specific settings and measurements before startup. Groups get together periodically and review the process safety dashboard to respond to risks, determine what tests to perform, and identify what would happen under certain conditions. Management gets automated reporting of process safety risk metrics and discuss how to respond and improve the situation.

The 2017 Nobel Prize in economics went to another behavioral economist, Richard Thaler, for his work in guiding positive rational decisions. He is credited with influencing public policy from automatic enrollment in retirement savings plans to organ donations, and smoking. When asked how he would spend the 9 million Swedish kronor ($1.1 million USD) he replied *"as irrationally as possible."* (Ben Leubsdorf)

CHAPTER 6

Decomposing Risk

Prior to the space shuttle Challenger disaster, NASA routinely estimated the risk of a catastrophic event at 1 in 100,000 (an event which if they conducted a launch every day, would occur only once in 274 years). They used a subjective scoring risk matrix approach to determine this estimate.

Risk matrices are commonly used in the chemical industries, but are a crude and sometimes misleading way to measure risk. The tool is not used by insurance actuaries because it has no statistical validity. As a quantitative measurement tool, there is no assessment of uncertainty and no way to measure improvement.

One of the issues people frequently cite when measuring an event which has never occurred, is that there is no data to work with. Following the explosion of the space shuttle Challenger, NASA changed their estimates to be based on quantitative measurements from component tests. In the multi-decade history of rocket and shuttle launches, they had many part and process failures.

The prior malfunctions included the specific components which ended up failing in the Challenger launch. The O-rings of the booster rockets had shown on multiple previous launches to have integrity issues when launches occurred at low temperatures due to embrittlement. With near freezing temperatures at the time of launch, the Challenger faced a higher than normal level of risk. (Vaughan) Decomposing the component test data revised the estimate for disaster under normal conditions to 1 in 78 with a range of 1 in 168 to 1 in 36. (Broad)

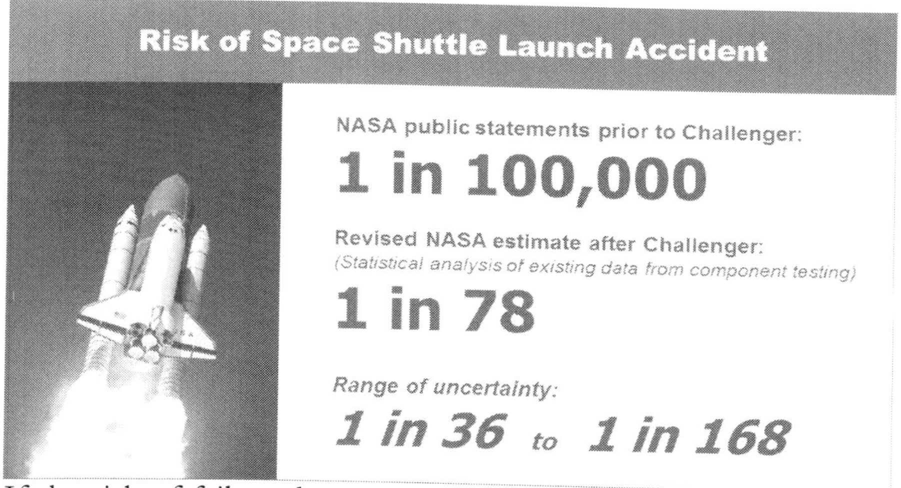

If the risk of failure does not meet certain standards, action may be taken, but this requires understanding risk through actual data measurement. Decomposing the question with component tests provided NASA a better assessment of risk.

Bayesian statistics uses expert estimates for building a model, but immediately checks and improves the model with real data measurements. NASA never did this until after the disaster and their risk matrix was wrong by many orders of magnitude.

CHAPTER 7

Problems with the Risk Matrix

The major problem in using risk matrices is in the belief that they actually measure risk. To measure risk the model would need to test the expert opinion with actual data. Without data, all you have is a theory, or an estimate based on assumptions. The quality of the assumptions depends on the experience of the expert using the matrix tool.

The risk matrix is a commonly used tool in the chemical industry to assess risk. The tool uses a "heat map" approach built by ranking severity on one axis and frequency on the other. Subject matter experts are asked to subjectively score the risks of event scenarios. These scores may be descriptive (low, medium, high) or numerical. In the matrix example below, the risk score is expressed in letters A to F, and in order of magnitude numbers. In order to get to a risk classification, the expert estimates the frequency of the scenario event based on experience, or sometimes on statistical data. They also predict the potential consequence as a possible or likely outcome of the scenario event.

Risk Matrix		Consequence				
		Catastrophic Many Fatalities	*Critical* 1 or Few Fatalities	*Moderate* Severe Injury	*Minor* Loss Time Injury	*Negligible* First Aid
Frequency	once/year	A 10	A 1	B 10^-1	D	E
	once/10 years	A 1	B 10^-1	C 10^-2	E	F
	once/100 years	B 10^-1	C 10^-2	D 10^-3	E	F
	once/1,000 years	C 10^-2	D 10^-3	E 10^-4	F	
	once/10,000 years	D 10^-3	E 10^-4	F 10^-5		

Risk matrices, based on assumptions, experience and generic data are suboptimal. Possible causes are sometimes ignored by making assumptions such as:

1. That the plant is well maintained is a precondition (not a variable).
2. That the plant is being run by well rested competent operators is a precondition (not a variable).

Using more data improves the focus on the real risks, and therefore helps to reduce incidents while at the same time avoiding the costs from a misallocation of resources. This will provide better safety at a lower cost. [1]

---

[1] The numbers in the matrix are the individual fatality risks of the process without safety measures (raw risk). Safety measures are added depending on the risk class (A,B,C,D,..) to ensure that a maximum tolerated risk level is not exceeded.

The advantage of the Risk Matrix method is anyone can understand and use it without knowledge of data science or statistics. The disadvantage is the result produced is not statistically relevant and may lead to misallocation of resources.

Risk matrices may actually increase the risk to an organization by providing a false sense of security. On the one side some actions will be very conservative (such as the Safety Integrity Level (SIL) of protective devices) while on the other side risks originating from lack of maintenance or from operating errors may not be adequately covered. A study of companies in the chemical industry found wide ranges in the risk scores produced by different SMEs ranking the same process. The study showed expert bias played a major role. (Lauridsen, 2002)

In addition to showing a wide range of estimating ability with subject matter experts, the risk matrix has several other inherent issues:

Range compression: Risks which estimators judge differently are measured identically. Experts must score a threat which they estimate to happen only once in 90 years in the same category as one that happens once in 10 years. Whereas, a risk estimated at once in 8 years is in a different category as one that happens once in 10 years. Small differences in subjective risk can lead to a different risk classification. Similarly, different judgement on the severity scale can increase the classification difference to 2 risk classes, or 2 orders of magnitude. The intention is to err on the safe side by overestimating risk, but in some cases will also go in the unsafe direction, which can lead to misallocation of resources.

Limited comparisons: Subject matter experts may only have expertise over a portion of the plant or a part of the possible causes. Various estimators are calibrated differently and scores are compared only to what they know. The resulting inaccuracies are from the lack of precision of the measurement instrument, the SME. The result can also depend on assumptions of frequency, which represent certain boundary conditions, like the presumed presence or absence of ignition sources.

Errors introduced in calculations: Even when used as designed the tool can actually rank riskier events lower than less risky events. This can produce results as Tony Cox, Massachusetts Institute of Technology PhD, stated, *"[they] can be worse than useless" leading to worse-than-random decisions."* (Cox)

Poor decision guidance: Reliance on these non-statistically based methods produces inaccurate results. Investments and resources may be allocated to the wrong areas, and the opportunity to make better fact-based decisions may be missed.

Monte Carlo Simulation

A statistically valid method to address the issues with risk matrices is the Monte Carlo Simulation. These models calculate the probability of an event based on data from various threats. The simulations use this data to determine the frequency of various events aligning at the same time. This is a commonly used tool by actuaries in the insurance industry to calculate risk.

Frequency: Risk varies over time due to many factors such as extreme weather, unplanned shutdowns, or automated controls being set to manual mode. Most major incidents happen when several of these factors occur at the same time. This is classified as the Swiss Cheese model and will be explored in the next chapter. Risk matrices ignore this situation and rank all risks independently, which adds to the flaws with the method.

With appropriate analytics, process safety experts are able to measure risk, and use foundational methods such as risk matrices, HAZOP studies, and Bowtie Analysis to determine appropriate countermeasures. Experts should frame the process to monitor rather than estimate the results. This leverages subject matter expertise to identify threats and potential consequences they pose to the organization. They bring context to the data which creates risk reduction with appropriate statistical monitoring.

CHAPTER 8

The Swiss Cheese Model

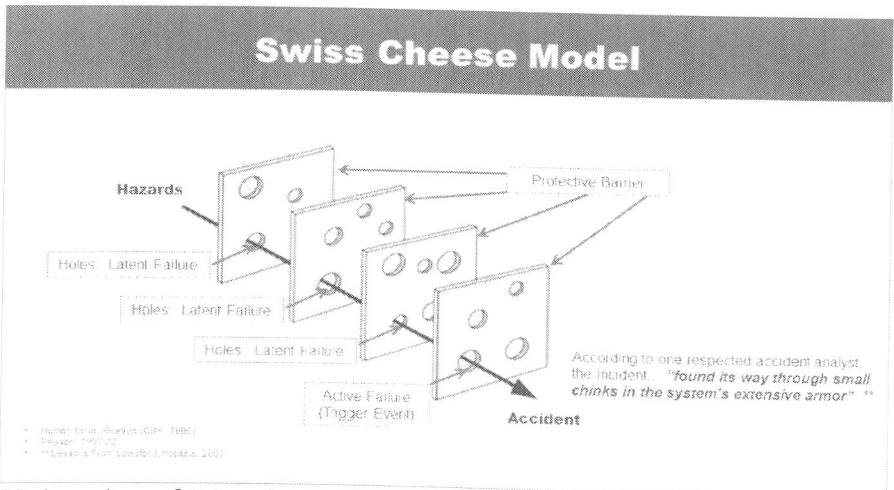

University of Manchester psychologist, James Reason developed the Swiss Cheese Model in 1990 to explain process safety disasters. Each major accident appeared as a singular event with a unique set of consequences. However Reason found they all occurred when a series Independent Protection Layers (IPL) developed holes which eventually coincided to allow for the potential of a catastrophic event. Usually holes occurred slowly through latent failures over time, long before being triggered by some action which started the incident.

Companies have used Hazop Studies, PHAs, and Bowtie diagrams to look for these holes. With new instrument and sensor technology, automated monitoring systems can track critical safety parameters along with recent audit data. Using an analytic framework, early warnings can be constructed to identify potential issues and provide prescriptive guidance. We can also track the corrective or preventative actions being taken for future learning and continuous improvement.

This "umbrella-like" framework offers a holistic approach to analytic monitoring which scans for potential pathogens that compromise overall process safety. We are now going beyond the IPL and creating a Learning Protection Layer (LPL), an early warning system which identifies weaknesses and continuously improves the process safety model.

Reason's Swiss Cheese Model presented two areas where protective systems break down: **Active failures and Latent conditions.**

Active failures are *"violations committed at the 'sharp end' of the system by pilots, air traffic controllers, police officers, insurance brokers, financial traders, ships' crews, control room operators, maintenance personnel and the like."*

"Like pathogens, **latent conditions** *such as poor design, gaps in supervision, undetected manufacturing defects or maintenance failures, unworkable procedures, clumsy automation, shortfalls in training, less than adequate tools and equipment – may be present for many years before they combine with local circumstances and active failures to penetrate the system's many layers of defenses."*

Although active failures are often the trigger event, Reason proposed they are rarely the cause, but usually the consequence. Reason and others found most process safety breaches occurred through both active and latent failures but they also may occur purely from latent conditions with no active failure.

There is a risk that "Failure Domains" (or holes in the protection layers) may occur even momentarily, and by themselves may not lead to an event. However in combination with other failure domains the level of risk rises significantly, opening up higher possibilities of a catastrophic event.

Reason argued that *"safety specialists could be directed more profitably towards the proactive identification and neutralization of latent failures."* This is where the LPL platform and Industry 4.0 excel.

Industry 4.0: Reviewing past disasters

Warning signs exist but are not recognized or appreciated prior to every major disaster. The following vignettes look at five past catastrophes, including upstream and downstream oil facilities, chemical, natural gas, and nuclear plants. A Swiss cheese diagram is provided for each, which identifies the major holes in the protection layers that made the event possible. Look for the critical parameters to monitor which could have prevented the disaster or minimized the consequences.

Texas City Refinery Explosion, 2005

The company operating the Texas City Refinery was actively seeking to become an HRO, and was presenting the site with a safety award the day of the disaster.

Operators at the refinery made a decision to restart after maintenance was performed, but communication was poor and management was unaware a startup was underway.

The operators began filling a 170 ft. tall distillation tower, designed to have a maximum liquid level of 9 ft., leaving the remaining 161 ft. for gases. This required proper management of inflow and outflow valves to maintain the level. Because downstream tank levels were high at the time of startup the operators decided not to open the outflow valve for the first three hours. Automatic controls would have kept the level at about 6.5ft., but the APC was turned off which is not uncommon during many plant startups. The tank level instrument showed the tower was at 9 ft., but this was its measurement limit, so it never read any higher. The backup measurement was a physical glass viewer which for years had been too weathered to view.

After several hours the measurement started showing the liquid level was decreasing to 8.5 ft. while the actual level was 158 ft. The tower temperature was also brought up faster than specified, and the expansion of liquid and gas triggered the pressure relief system, which vented a vapor cloud. The plant had proposed replacing the vent with a modern flare which burns off unwanted gases, but funding was never approved. Heavier than air vapors descended to the ground where an idling truck ignited the gases causing the explosion. A high level of casualties occurred due to an employee meeting in a nearby trailer office (not a blast resistant building). (Hopkins, Failure to Learn)

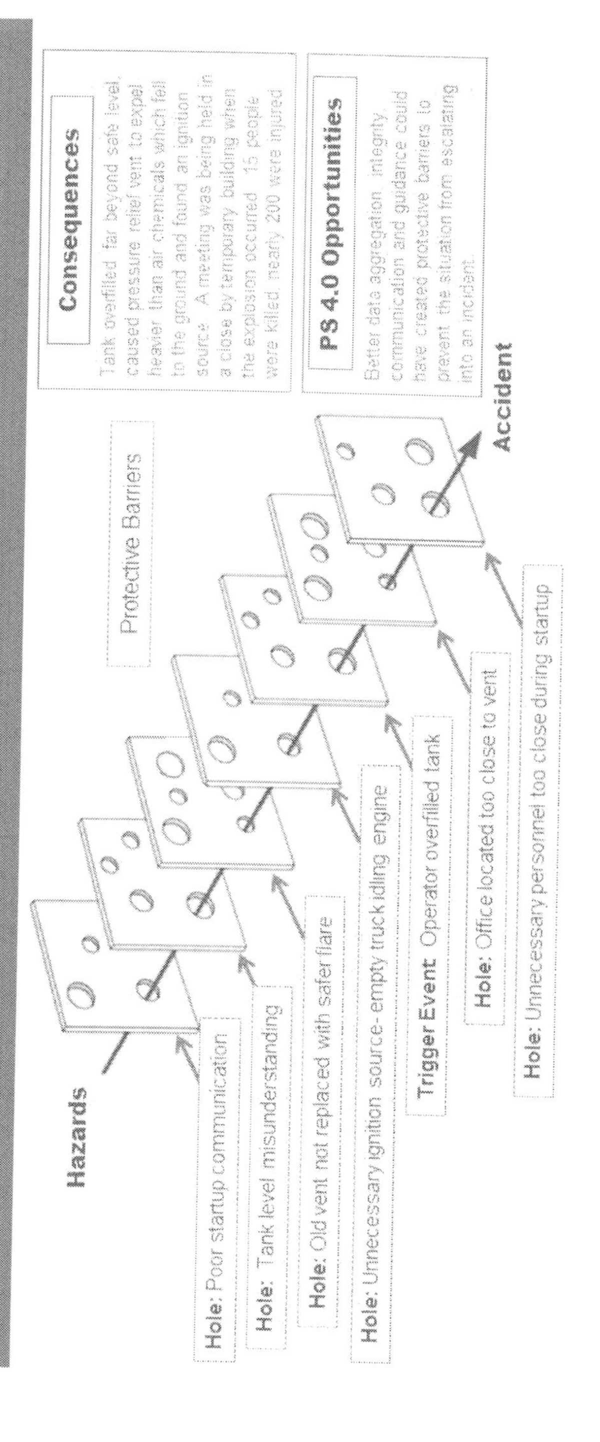
Texas City Refinery: Swiss Cheese Model
Hazards
Protective Barriers
Hole: Poor startup communication
Hole: Tank level misunderstanding
Hole: Old vent not replaced with safer flare
Hole: Unnecessary ignition source- empty truck idling engine
Trigger Event: Operator overfilled tank
Hole: Office located too close to vent
Hole: Unnecessary personnel too close during startup
Accident
Consequences
Tank overfilled far beyond safe level, caused pressure relief vent to expel heavier than air chemicals which fell to the ground and found an ignition source. A meeting was being held in a close by temporary building when the explosion occurred. 15 people were killed, nearly 200 were injured
PS 4.0 Opportunities
Better data aggregation, integrity, communication and guidance could have created protective barriers to prevent the situation from escalating into an incident.

Institute Pesticide Explosion, 2008

This explosion occurred in a crop science plant located in Institute, West Virginia, where the company manufactured Methomyl. There had been a previous disaster in 1993 when an explosion killed one employee and injured two others. Management control over the critical process parameters had never been fully corrected and similar issues led to both events.

In 2005 OSHA inspected the plant and cited "deficient process hazard analyses," but never followed up to ensure these issues had been corrected.

Three weeks before the 2008 explosion, 48 hazards were identified during a safety audit, which were supposed to have been corrected but were not. The unit only had one technical advisor who in the days leading up to the explosion had worked 15 to 17 hours per day.

On the day of the explosion operators with inadequate training (according to the CSB), skipped comprehensive pre-startup procedures including performing equipment calibration, checking valve positions, and completing all the process steps.

The production process involved reactions with a number of very hazardous, flammable and explosive chemicals. The combination of improperly adjusted equipment and incorrect valve positions led to an uncontrolled chemical reaction with extreme heat and pressure. This caused a 4,500 gallon vessel to explode and start an intense fireball. Emergency responders were unaware of the chemical risks they faced and were injured in the relief effort. Two people were killed, and 16 injured (6 were first responmders). (CSB, Report)

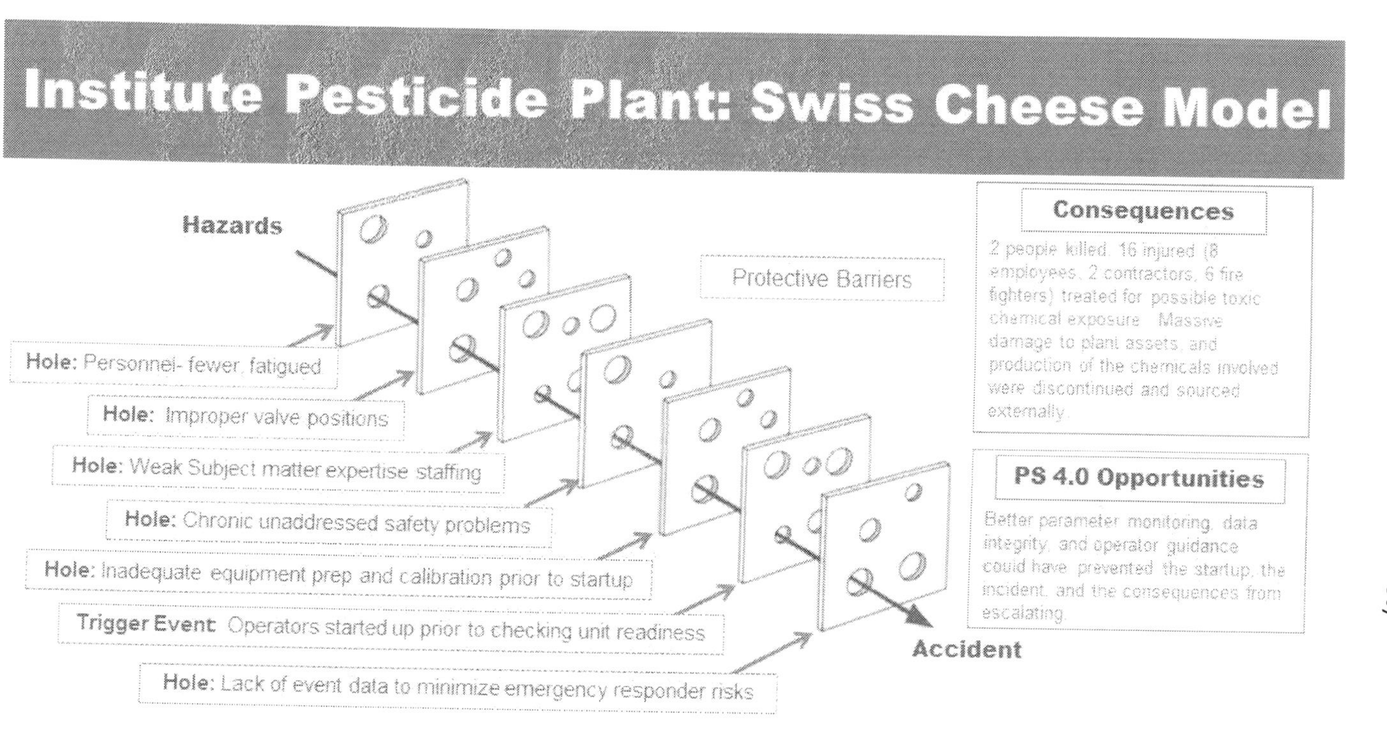
Institute Pesticide Plant: Swiss Cheese Model
Hazards
Protective Barriers
Consequences
2 people killed, 16 injured (8 employees, 2 contractors, 6 fire fighters) treated for possible toxic chemical exposure. Massive damage to plant assets, and production of the chemicals involved were discontinued and sourced externally.
PS 4.0 Opportunities
Better parameter monitoring, data integrity, and operator guidance could have prevented the startup, the incident, and the consequences from escalating.
Hole: Personnel- fewer, fatigued
Hole: Improper valve positions
Hole: Weak Subject matter expertise staffing
Hole: Chronic unaddressed safety problems
Hole: Inadequate equipment prep and calibration prior to startup
Trigger Event: Operators started up prior to checking unit readiness
Hole: Lack of event data to minimize emergency responder risks
Accident

Deep Water Horizon, 2010

In the weeks prior to the Deep Water Horizon explosion a confidential survey found that workers were fearful of reprisal for reporting safety problems to the company, indicating a weak HRO culture.

A contractor had just completed installing a cement casing on the well head of this ultra-deep water off shore drilling rig. An internal document shows that experts were skeptical whether this casing would work. A special nitrogen foamed cement plug had not yet been installed in the cement casing. Prior to drilling, a negative pressure test was performed, but there was no agreed level or procedure for what would be an acceptable result, and the measurements were misinterpreted as adequate.

Measurement instruments indicated gas bubbling in the well which can be a sign of an impending blowout, but no action was taken. Rising reservoir fluid should have been detected by the water inflow and mud outflow ratio but no adequate measurement was being tracked.

When the hydrocarbons reached the drilling rig's floor, the ship's diesel generators acted as the ignition source for the explosion and fire.

Although the wellhead was equipped with a Blowout Preventer (BOP), it was not fit for service. The solenoid was wired improperly, and a critical battery was drained. The status of this safety critical system was unknown as no monitoring was in place. The BOP failed and oil continued to spill from the well for nearly three months when finally a permanent cap was deployed. 11 people were killed, 17 injured, and costs estimated at $54 Billion. (CSB BOP Failure Deepwater Horizon)

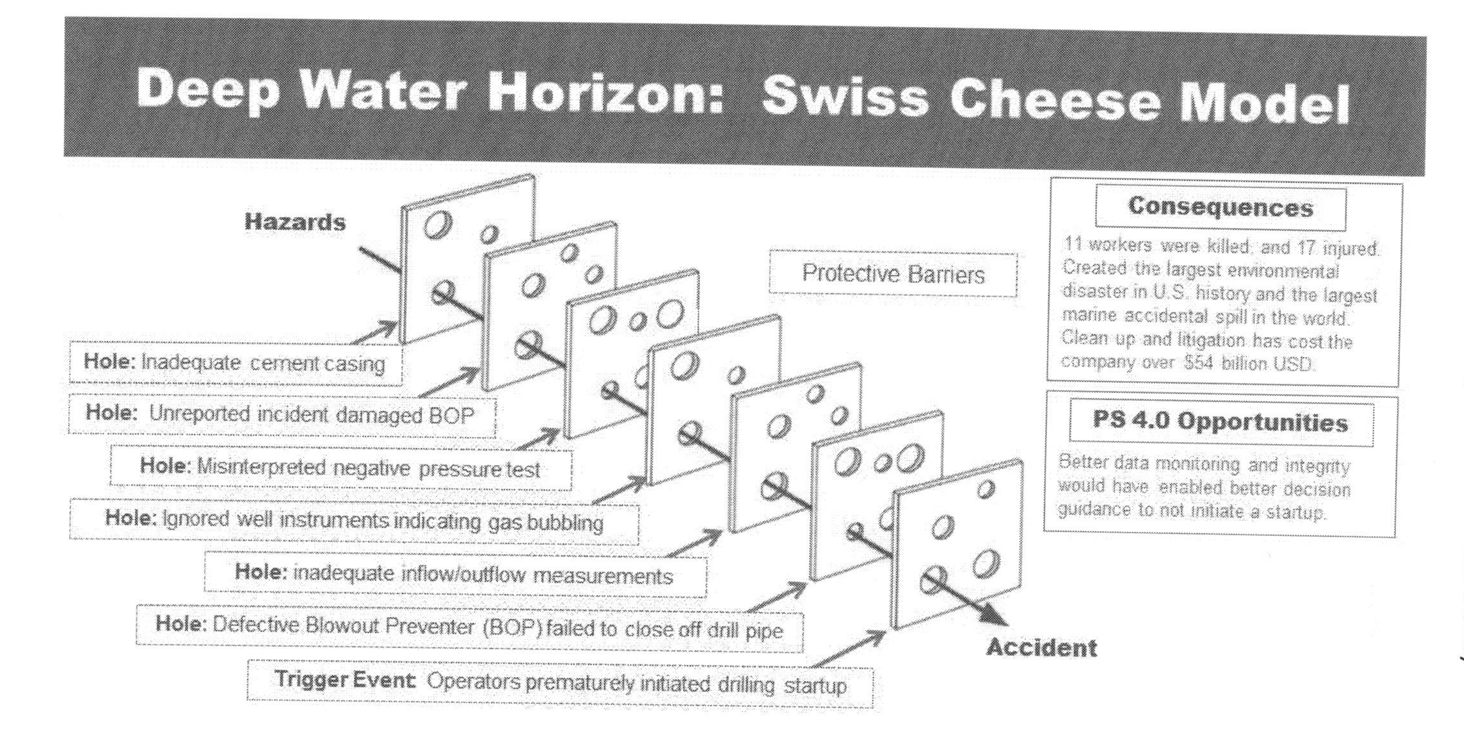
Deep Water Horizon: Swiss Cheese Model
Hazards
Protective Barriers
Consequences
11 workers were killed, and 17 injured. Created the largest environmental disaster in U.S. history and the largest marine accidental spill in the world. Clean up and litigation has cost the company over $54 billion USD.
PS 4.0 Opportunities
Better data monitoring and integrity would have enabled better decision guidance to not initiate a startup.
Hole: Inadequate cement casing
Hole: Unreported incident damaged BOP
Hole: Misinterpreted negative pressure test
Hole: Ignored well instruments indicating gas bubbling
Hole: inadequate inflow/outflow measurements
Hole: Defective Blowout Preventer (BOP) failed to close off drill pipe
Trigger Event: Operators prematurely initiated drilling startup
Accident

Longford Natural Gas Disaster, 1998

Operators consistently faced an overload of noisy information with about 8,500 alarms triggering in a 12-hour shift. The condensate transfer system was frequently run past its measurement limit. The night before the incident, condensate transfer was run for hours causing the warm oil pumps to shut down.

In a restructuring plan, engineers had been removed from the natural gas plant in 1992 and relocated to a central office in Melbourne. The process plan was to have these subject matter experts remotely monitor temperatures, pressures, condensate levels, and other parameters, but much of this information was recorded on paper and frequently not sent to Melbourne. 30% of these tracking processes were not being executed on the day of the incident.

The physical isolation of engineers from the operations staff in theory would not inhibit communication as they were only a phone call apart, but in practice this was not the case.

Communication during shift changes was also inadequate and the day of the incident the new shift operator closed a valve which raised condensate levels even higher, creating the opportunity for the accident sequence.

Very low temperatures in the heat exchanger caused the metal to contract. Joints and bolts no longer fit snugly and leaks occurred which could not be fixed when maintenance attempted to tightened them. Operators sent warm oil flow into the freezing cold heat exchanger which exploded due to cold metal embrittlement, a safety condition of which the operators were unaware. (Hopkins, Lessons From Longford)

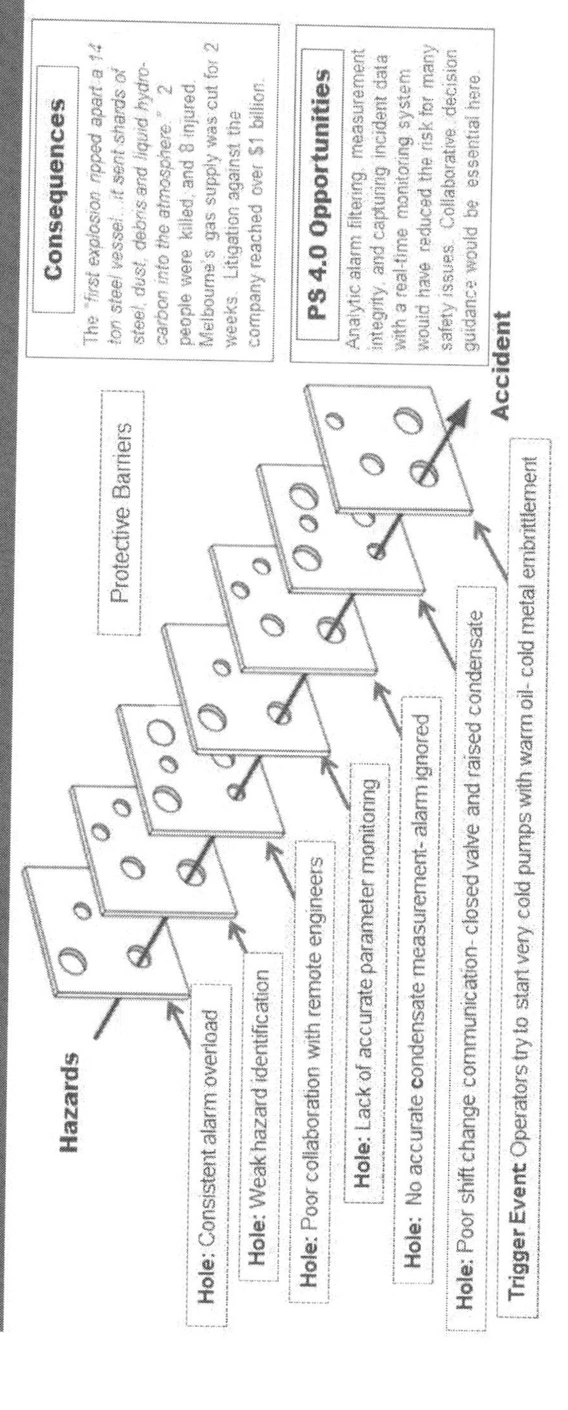
Longford Gas Plant: Swiss Cheese Model
Hazards
Protective Barriers
Hole: Consistent alarm overload
Hole: Weak hazard identification
Hole: Poor collaboration with remote engineers
Hole: Lack of accurate parameter monitoring
Hole: No accurate condensate measurement- alarm ignored
Hole: Poor shift change communication- closed valve and raised condensate
Trigger Event: Operators try to start very cold pumps with warm oil- cold metal embrittlement
Accident
Consequences
The "first explosion ripped apart a 14 ton steel vessel…it sent shards of steel, dust, debris and liquid hydro-carbon into the atmosphere." 2 people were killed, and 8 injured. Melbourne's gas supply was cut for 2 weeks. Litigation against the company reached over $1 billion.
PS 4.0 Opportunities
Analytic alarm filtering, measurement integrity, and capturing incident data with a real-time monitoring system would have reduced the risk for many safety issues. Collaborative decision guidance would be essential here.

Three Mile Island Disaster, 1979

After an unsuccessful attempt to clear a blockage in the secondary water line filters using compressed air, operators tried applying compressed air directly to the water. A stuck open check valve allowed water to enter the instrument air lines, which caused a series of pumps to turn off.

Without the cooling process, the nuclear chain reaction system automatically tripped a shut down. The auxiliary pumps automatically started but were unable to pump water because the valves had been closed for maintenance.

The control panel misled operators as the system showed the pressure relief valve was closed when it was stuck open. The valve position could have been verified by a temperature indicator downstream, but the indicator was located on the back of the seven foot high instrument panel they were facing, and they were not trained to check this.

Hours passed before the next shift arrived and were able to diagnose the actual problem. Operators had no indication of the coolant water level but were concerned about overfilling and turned off the emergency core cooling pumps.

A series of reactions caused the nuclear fuel rod cladding to melt, and a small explosion to occur.

32,000 gallons of radioactive coolant escaped from the primary loop into the general containment building. Although no one was killed, the incident caused the downfall of the nuclear power industry in the United States.

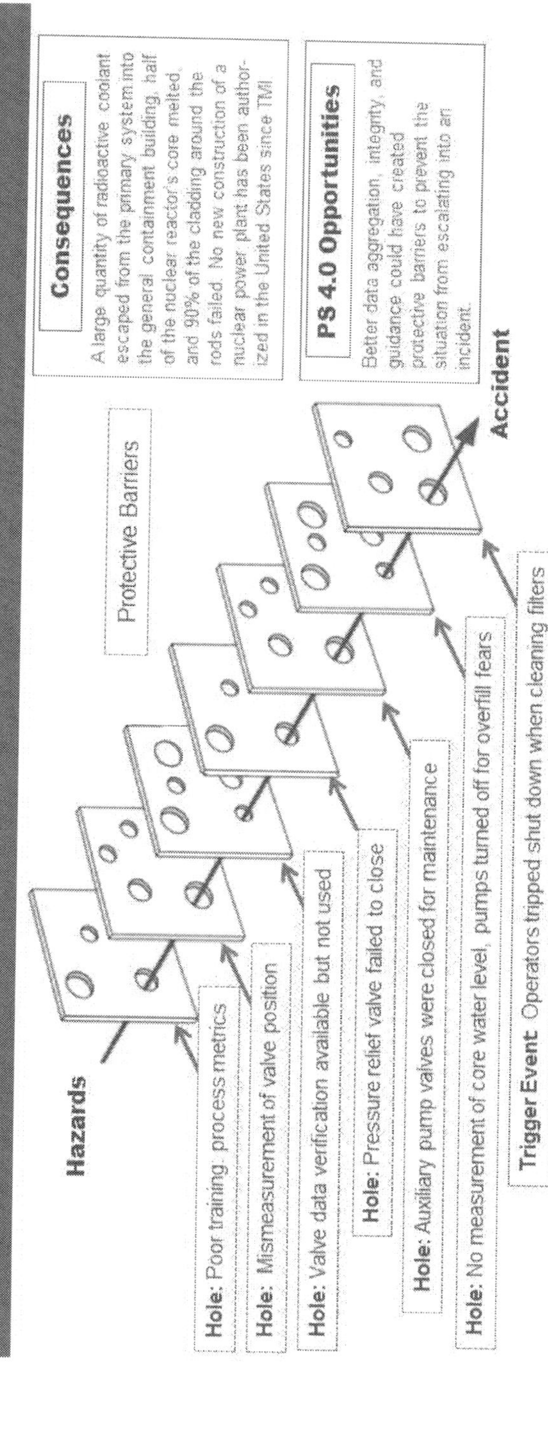
Three Mile Island: Swiss Cheese Model
Hazards
Protective Barriers
Accident
Hole: Poor training, process metrics
Hole: Mismeasurement of valve position
Hole: Valve data verification available but not used
Hole: Pressure relief valve failed to close
Hole: Auxiliary pump valves were closed for maintenance
Hole: No measurement of core water level, pumps turned off for overfill fears
Trigger Event: Operators tripped shut down when cleaning filters
Consequences
A large quantity of radioactive coolant escaped from the primary system into the general containment building, half of the nuclear reactor's core melted, and 90% of the cladding around the rods failed. No new construction of a nuclear power plant has been authorized in the United States since TMI
PS 4.0 Opportunities
Better data aggregation, integrity, and guidance could have created protective barriers to prevent the situation from escalating into an incident.

CHAPTER 9

Bowtie Analysis with Industry 4.0

Plant operators are focused on maximizing production. Operators are also responsible to monitor process safety, but believe that process engineers and safety instrumented systems *(SIS)* have this handled. By HRO standards this is not very mindful.

A process safety platform should keep all key stakeholders informed of both long-term drifts and up to the minute situations. Bowtie diagrams are useful for identifying independent protections, from automated shut downs to what Industry 4.0 calls human/instrument interactions, (Industrial Internet of People, or IIoP).

In some cases, threats may have been identified and forgotten by the organization years ago. This is a risk for very slow-moving threats which may drift over many shift changes, months or even years. Analytic alarms (as opposed to typical fixed set-point alarms) monitor based on statistical probability.

The Bowtie method focuses on putting both barriers in place to prevent an event, and barriers to mitigate the potential consequences once an event occurs. The barriers are IPLs.

A Learning Protection Layer *(LPL)* may trigger a model based alarm from relationships between multiple parameters and provide a longer advanced warning system. The LPL is more focused on maintaining process safety health, than crisis management.

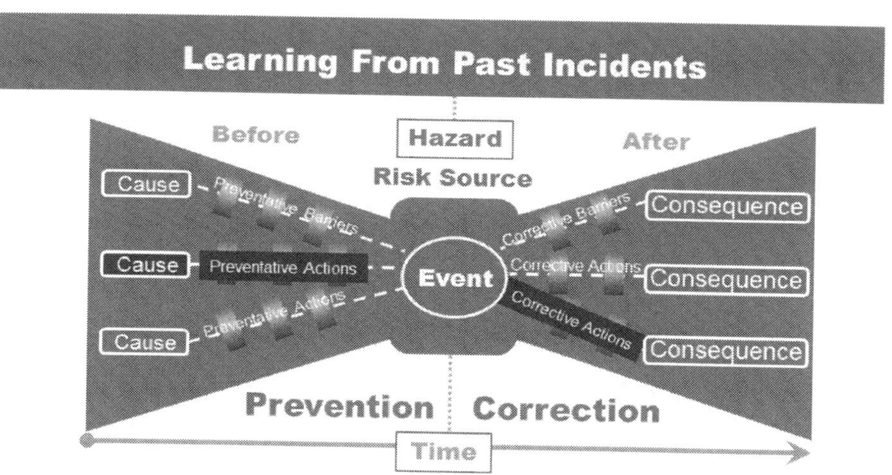

Tracking Causes and Actions: When threats are identified, the system allows users to track the assigned cause and action that was taken to either prevent an event or a consequence from an event. This builds a database of knowledge about safety which can be analyzed with paretos and other tools to determine the best future decisions.

Industry 4.0 enables analytic monitoring systems which continually improve over time, as they reduce variation by detecting statistical anomalies. These system track specific events and log the preventative actions, corrective actions, and assignable causes attached with the data. These systems may also incorporate user role management with revision rights for some personnel and read only for others.

These records may be reviewed with Pareto analysis, for continuous improvement. A collaboration platform may tie the safety information together in a knowledge base which can be accessed by all stakeholders.

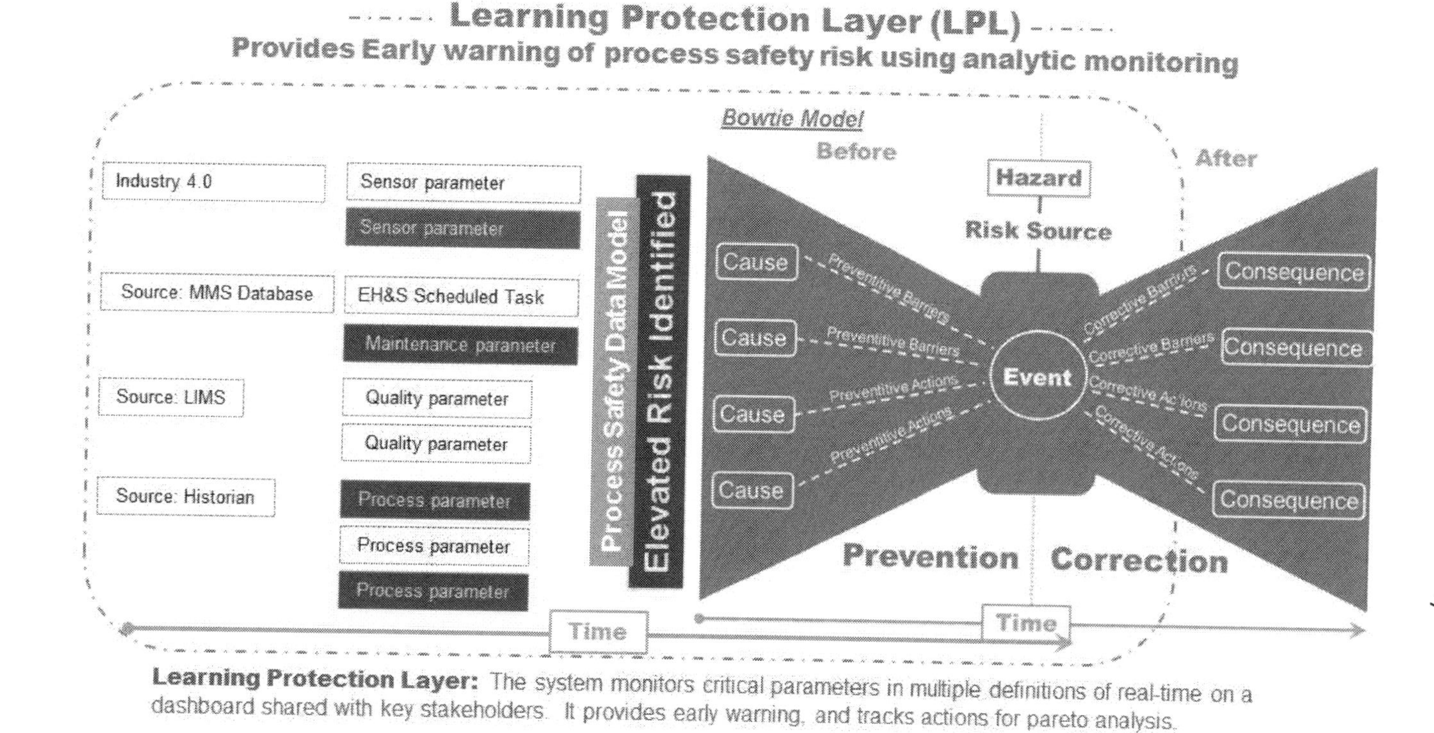

Learning Protection Layer: The system monitors critical parameters in multiple definitions of real-time on a dashboard shared with key stakeholders. It provides early warning, and tracks actions for pareto analysis.

CHAPTER 10

Process Safety Design 4.0

The fourth revolutionary wave in manufacturing is Industry 4.0, where processes move beyond computer automation to an integrated analytical platform.

During the 3rd wave, a wide variety of systems were independently designed which created many very different data silos. These included process historians, laboratory information management

Systems (LIMS), distributed control systems (DCS), manufacturing execution systems (MES), Enterprise Resource Planning (ERP), maintenance management, and many other independent databases which may include audit and action point tracking, near miss and incident tracking, training completion, number of people present in the plant, presence of high risk work in the plant, weather, etc.

With Industry 4.0 we integrate all of these current and future data sources to create a protective umbrella type architecture. The overall picture of the operation allows analytic monitoring to identify early warnings of potential threats which can be quickly remediated. This also provides a collaborative environment for key stakeholders to see how process safety is performing at any point in time. This is important as lack of communication was a critical hole in many past disasters.

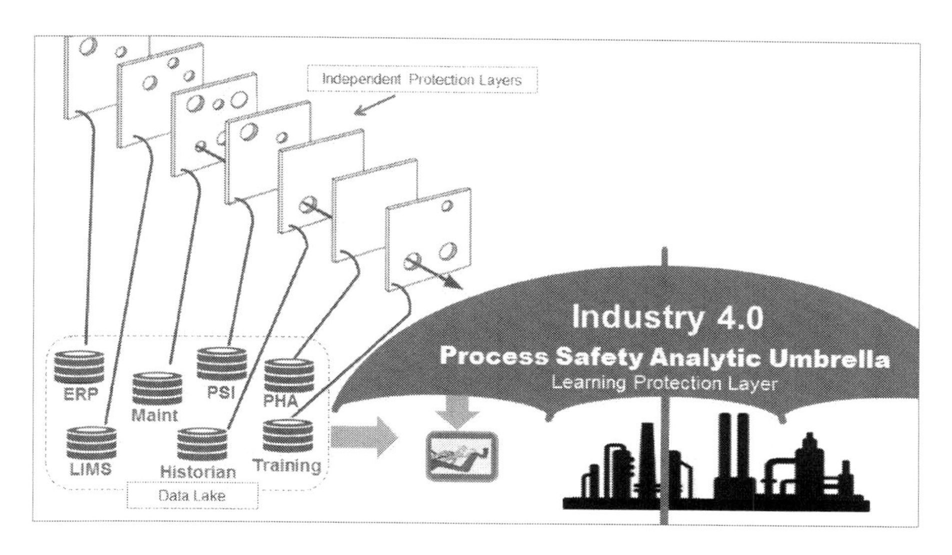

Industry 4.0: Four Basic Design Principles

1. Interoperability:

Interoperability is the relationship between devices such as sensors, instruments and machines with people. This is the ability to leverage all of the available data from many sources to drive better decisions *(IIoP / IIoT).**

2. Information transparency:

This refers to aggregating data from a wide array of sources to enrich more effective modeling. Information transparency allows data to be aligned and contextualized in a more holistic approach, which can better represent the process.

3. Technical assistance:

Technical assistance has two aspects:

- The first is to provide a system which supports human decision making with real-time rationalized data to solve urgent problems.
- The use of robotics under human control (cobotics) to support human tasks which are physically too unsafe or unpleasant for human workers.

4. Decentralized decisions:

Where possible, analytics should be designed into instruments and robotics to allow decentralized independent decisions so tasks can be completed autonomously. *(Edge analytics)*

* *IIoP- Industrial Internet of People*
IIoT- Industrial Internet of Things

(Hermann, Pentek and Otto)

Process safety is well suited to leverage the design principles of Industry 4.0. The holistic structure addresses the issues found in the Swiss Cheese model and will reduce the risk of major incidents.

Inline instruments used to require special housing in stable environments with climate controlled sheds. These buildings could cost more than $1 million. Today inexpensive sensors may capture many variables at the same time and wirelessly transmit results without the need for instrument sheds. This is developing more available low cost data for an Industry 4.0 platform.

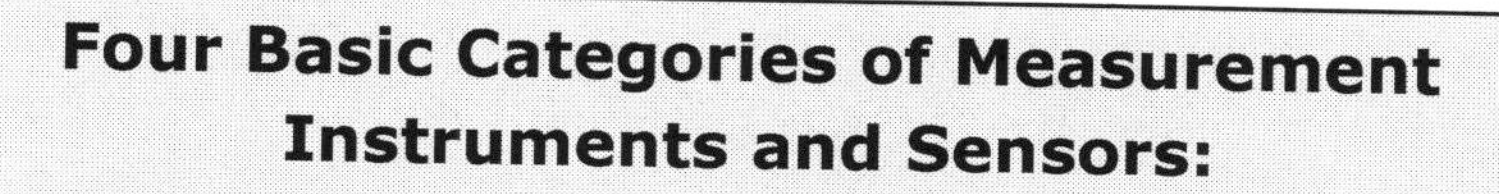

Four Basic Categories of Measurement Instruments and Sensors:

Inline: The measurement sensor is integrated into the process and constantly monitors materials as they pass by.

Online: Material is sampled, analyzed and returned to the process.

At line: A sample is removed from the process and tested on location.

Off line: A portion is removed from the process, and taken away for measurement usually in a central location such as a lab.

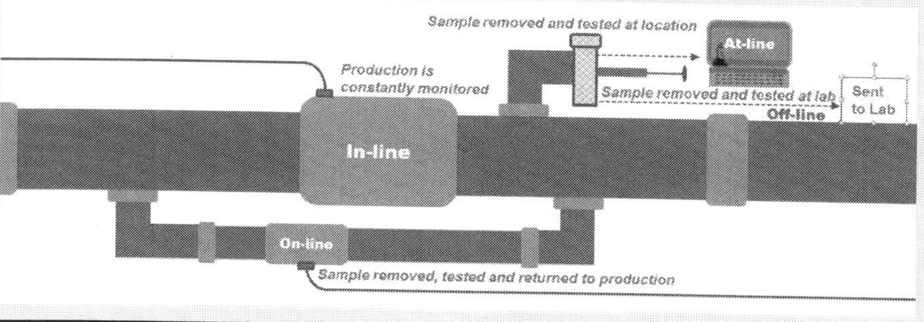

The information from these measurement systems could end up being stored in a variety of different databases.

Aggregating this information together is an essential part of Industry 4.0 design as the combination provides more complete context of the process. Aggregation needs to account for lead/lag relationships to provide correct data alignment. Monitoring in real time is essential to support urgent decision making.

Human controlled robots (cobots) are able to capture data and make maintenance repairs far faster and safer than in the past. They are being sent to locations hostile to humans, such as towers and furnaces, to do inspections and repairs. The obvious advantage is that they can be piloted by someone sitting safely behind a computer away from dangerous gases, temperatures and spaces.

CHAPTER 11

Use Cases

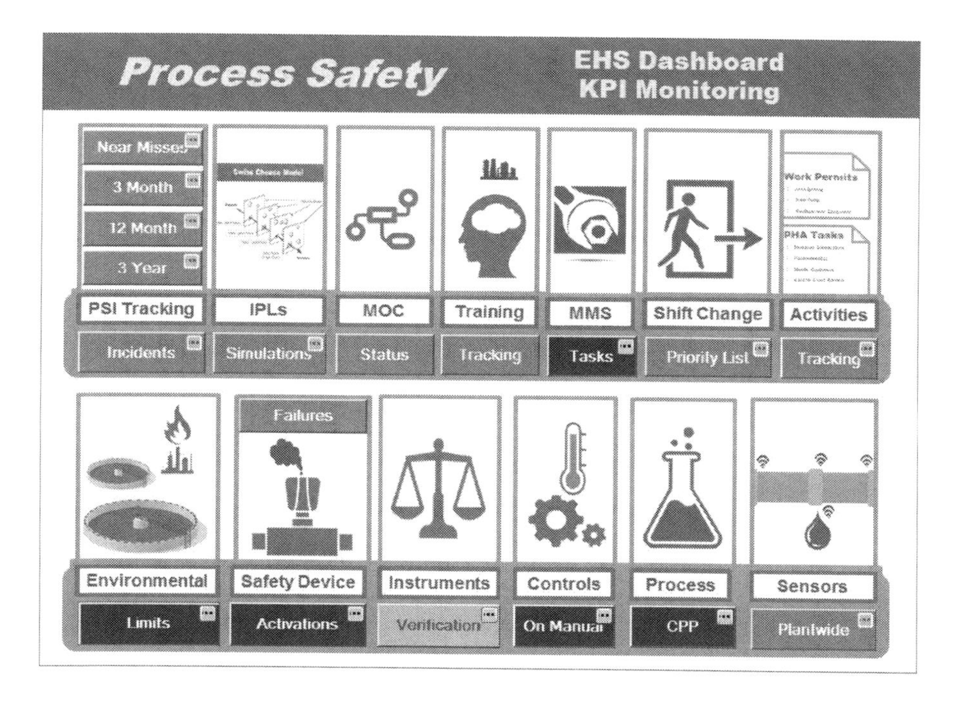

Aggregated real-time analysis enhances process safety initiatives as they combine parameters to monitor factors such as corrosion, vibration, weather, energy consumption, and other factors.

There are significant process parameters which can help companies improve threat identification, including: pump temperature, fouling, valve positioning, torque, pressure drops, compressor energy usage, relief device triggering, and plant trips. If parameters like these are monitored in context, early warnings are manageable.

Monitoring safety threats Here are some examples from case studies of real-time monitoring projects which are catching potential threats before they develop.

Startup optimization: Historical data from previous startups can provide better guidance. Usually Advanced Process Control (APC) is turned off during startups, but the data from previous startups can be used to calculate decision guidance for correctly ramping up specific parameters.

Using data from even six previous good startups can provide statistical information for guiding startup parameters such as temperature and pressure. Operators can see the statistically significant range for high and low variation at each minute. This is referred to as a golden corridor (or golden tunnel).

Unusual measurements: When plant compressors are running unusually hard, where is all the pressure going? This may be an indication of a Loss of Containment (LOC).

One major chemical company began an initiative to break down and monitor energy consumption. Compressors were the third largest consumer of power at the site. They were able to detect statistically unusual spikes in compressor utilization while pressure remained normal, which helped detect incorrect valve positions. In one case, they found perfectly good product was being vented to the flare.

Improper valve positions have been the cause of many safety incidents. Another cause could have been a LOC due to corrosion causing leaks (another common safety hazard).

> **Example:**
>
> Calculated metrics are less obvious than just monitoring pressures and temperatures, and industry 4.0 offers the combined look as an evaluation.
>
> For instance, a pressure drop increases on a fouling heat exchanger, while at the same time the heat transfer trends downwards. The electricity consumption of the pump through the exchanger shows an increase due to higher viscosity from polymerization. The combination of factors triggers an alarm, alerting that something is headed in the wrong direction, and may lead to an incident.
>
> Classical process control would not provide this kind of early warning.

This type of automated monitoring usually results in multiple benefits to the company, including: improved process safety, lower energy consumption, increased process efficiency, strict environmental compliance, reduced maintenance costs, and on time delivery.

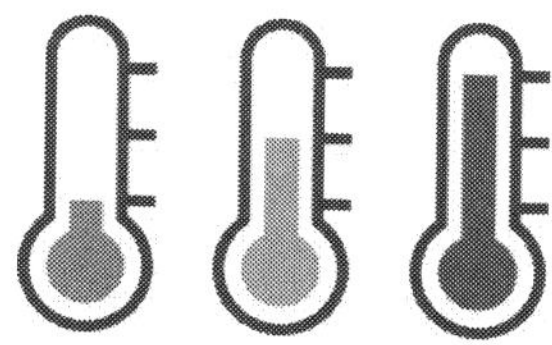

Pump maintenance: Analytic monitoring enables maintenance plans based on appropriate need rather than fixed schedules. For instance pump maintenance may be based on usage, temperatures, filters, etc. If feed stocks are unusually dirty, maintenance may be required sooner than would normally be scheduled. The opposite can also be the case, resulting in pumps taken out of service too often. An overheated pump which burns out creates an unplanned shut down, stopping production. This will impact plant profits with examples of this preventable instance costing $100,000 or more to perform a replacement. Often uncounted are the increasing process **safety risks, associated with unplanned startups.**

Tank levels: Chemical tank (or container) management is frequently involved in process safety incidents. Movement happens slowly, but complex relationships usually exist between multiple tanks and valve positions. While operator attention is focused on other tasks, situations can quietly escalate.

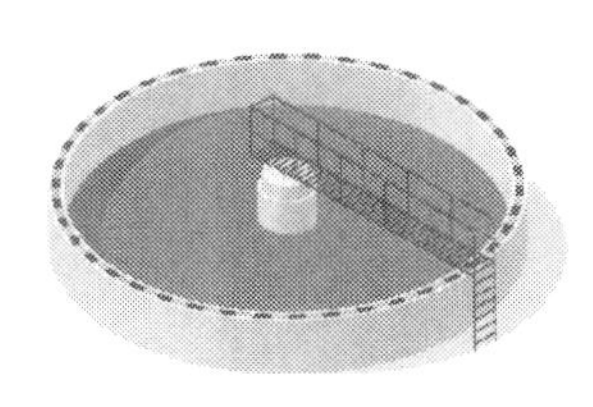

Analytic monitoring can give early warning when deviation from predicted levels arises. Changes in tank levels over time can be anticipated based on the relationship between flow rates and valve positions. These kinds of mismatches can be caught by an Industry 4.0 platform. In one case, prior to the new system being deployed, an undetected open valve caused product to pour onto the ground for six hours before being noticed and corrected.

In another case tank levels were dropping, but doing so slowly. Alarms would sound indicating a lower threshold, but operators just acknowledged them as there was no concern. After six shift changes the tank levels had dropped below the minimum, and damage occurred. Communication between shift changes was faulted for the mishap. This could have been prevented with a system detecting statistical anomalies and a collaborative platform where acknowledged alarms do not just disappear.

Weather extremes: Extreme weather conditions frequently increase the risk level at a process manufacturing plant. If tank levels are high in the effluent plant, excess rain can cause the system to overload, and sometimes require the plant to shut down or threaten LOC. Cold metal embrittlement caused the Longford gas explosion and the Challenger space shuttle disaster. Excess heat may cause overheating pumps, generators, or compressors. Monitoring early warnings of these conditions and taking measures to minimize the impact is critical. This may mean prioritizing tasks which are not normally a focus of attention such as bringing down tank levels in the effluent plant.

Monitoring measurements: In all of the case studies analyzed with the Swiss Cheese Model in Chapter 8, measurement integrity was a major factor leading to the incident. Operations personnel were making decisions based on missing or inaccurate data. In some cases there was a lack of trust in the data. Validating the data to assure it can be relied upon is essential for any Industry 4.0 project. Properly organizing the results in context creates a system which can monitor itself for data integrity.

Comparing parameters can help identify when unusual variation is occurring. This approach provides an early warning that either instruments are malfunctioning or there is a process safety issue. The goal is to provide quick detection of an LOC incident or a data integrity issue, which requires attention.

There are frequently secondary measurements which should match the primary source; if they are not consistent there is a data integrity issue. For instance valve positions can be checked against a material flow calculation. This type of discrepancy will be rare, but an analytic monitoring system can spot drifts early and prevent measurement error being the cause of poor decision making.

Automatically verifying instruments: Labs check their instruments and procedures for precision and accuracy, but this is usually a manual process requiring the lab technician to run a set of standards and review (sometimes hundreds of) charts. Some inline and online analyzers can be checked using known standards, or they can be compared with the lab. This entire process is an Industry 4.0 opportunity as all of the data can be monitored automatically, as the system will highlight only the issues which require review.

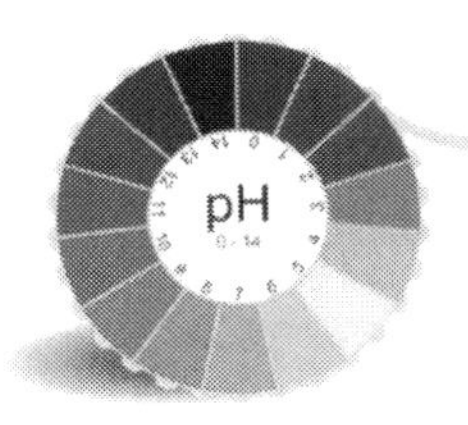

Corrosion: Coastal salt air, some chemicals, and low pH are highly corrosive to most pipes; these pipes in turn become a primary source for leaks. Monitoring corrosion is a challenge. For instance, if pH is not a priority for operations and not monitored closely, pipes may be stressed and susceptible to LOC incidents. In some cases pH levels may drop to zero or negative and can contribute to serious issues.

CHAPTER 12

Analytic Monitoring

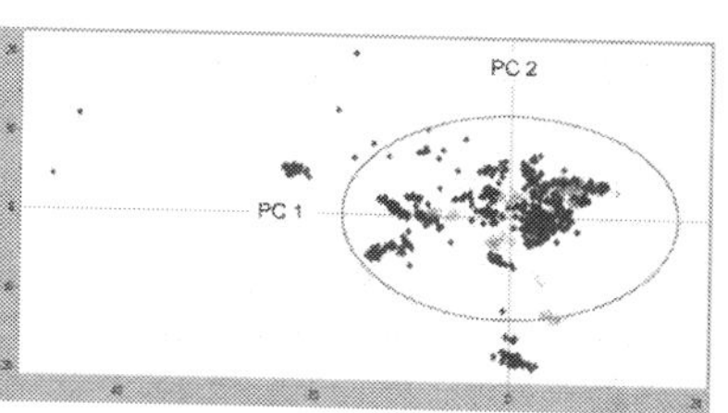

Analytic Models: Monitoring a multivariate model allows for early detection of potential issues as large combinations of factors are constantly reviewed to detect unusual variation. Principle Component Analysis (PCA) and Partial Least Squares (PLS) are the two most common analytic model approaches being used in the process industries.

PCA is employed to track adherence to steady-state or to identify the cause of unusual process behavior. PLS is tied to predicting continuous properties such as expected yield. Both algorithms have the ability to be used as a monitoring system and an investigative tool. Effective monitoring requires simplified decision guidance. If a data scientist is required to determine the next action, modeling will likely fail as a monitoring tool.

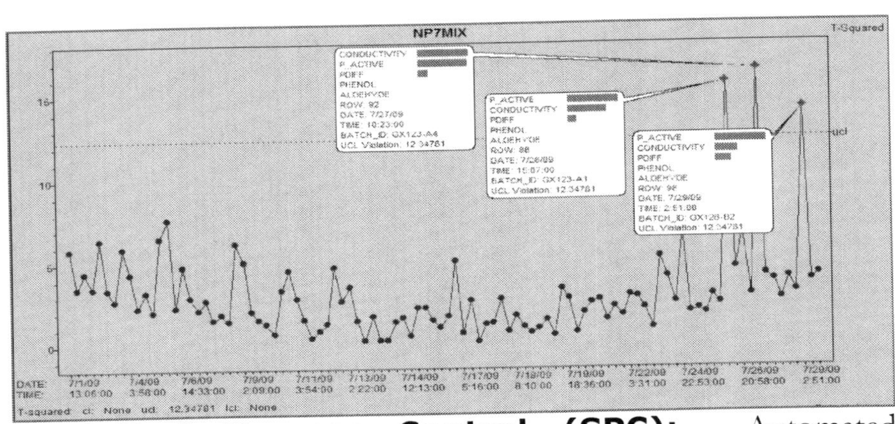

Statistical Process Control (SPC): Automated monitoring of SPC uses a proven methodology for reducing variation. This is usually applied for process optimization, but also applies to process safety as these projects frequently reduce the number of unplanned shutdowns. The analysis monitors for control, and statistically relevant pattern rules to provide early warning that the process may be headed out of specification. Operators can use these warnings to respond before the process wanders too far. This is also used for guidance to not take action when everything is running under control (common at shift changes).

Golden Tunnel: The Golden Tunnel is frequently used with nuclear reactors, but is uncommon in most other process industries. This is an excellent process safety tool, as the primary function is to aid in properly controlling startups.

Control limits are calculated (SPC), for each point in time throughout the startup based on several good start ups in the past. Automated controls are off during startups and the golden tunnel provides operator decision guidance during this critical time. This can be very useful for parameters such as temperature and pressure. One of the factors in the Texas City Refinery disaster was that the temperature in the distillation tower was brought up too quickly; this deviation would have been obvious had the data been compared to the previous experience set.

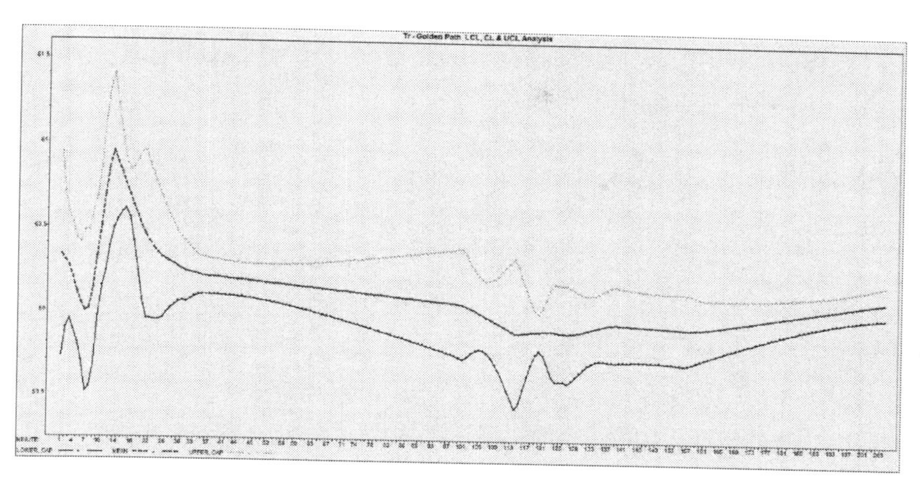

Simulations: Predictive simulations are used to develop process models which analyze the chemistry of current feed stocks and provide expected outputs during startups. If reality does not match predicted levels, key stakeholders are notified and resources are engaged to understand the discrepancy.

A Monte Carlo simulation runs repeated random sampling using an algorithm to determine the probability of a "Black Swan" event. If the critical data parameters change, a properly-constructed simulation will alarm when sub-optimal conditions exist. The process combines the probabilities of all excursions from normal to determine the likelihood of an incident, and the frequency of a Swiss Cheese Model occurance. Companies today are able continually monitor Monte Carlo simulations to understand when risk is elevated.

Bathtub Curve Hazard Distribution: The Bathtub Curve shows a typical failure distribution with initial issues which typically happen during the startup and commissioning phase, followed by a stable phase where failures are minimized, followed by a phase with an increasing number of failures due to normal wear through aging.

The failures may come from a variety of sources such as fouling, unplanned shutdowns, or relief valve events. Critical equipment is brought down for scheduled maintenance. Ironically, the recently serviced item will sometimes experience an early failure while it might have run much longer without preventative maintenance. Monitoring with Weibull Analysis can determine equipment which may be better serviced when needed rather than using a specified number of hours.

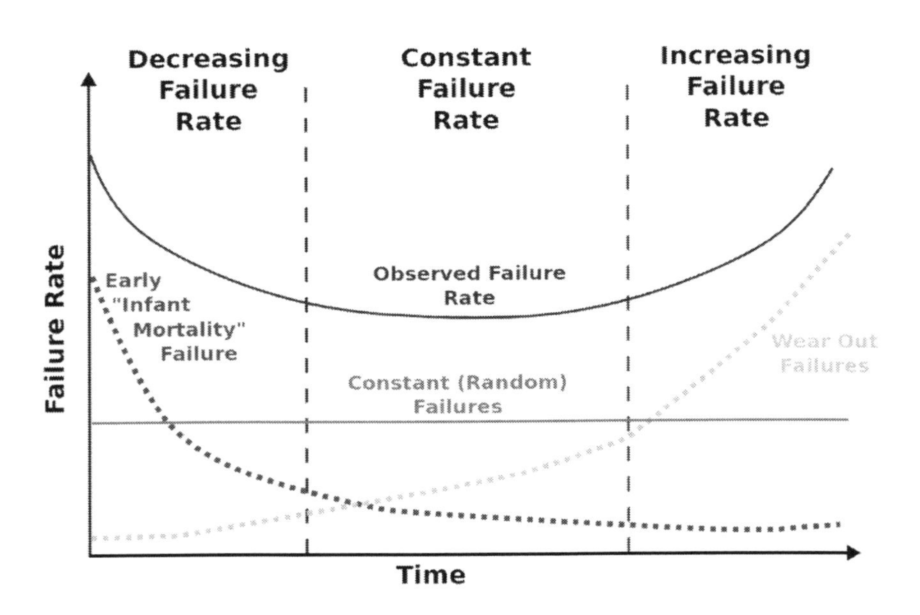

Weibull Analysis: This statistical function provides a view of the useful life of equipment. This monitors the stage of the Bathtub Curve through statistical analysis. This approach provides several useful functions for monitoring:

1. Over time a company will experience a range of Mean Time Between Failure (MTBF). This can be tracked and trigger decision points if the range is moving to shorter life.

2. If the ß (Beta) is **less than 1** then there is a higher distribution of infant mortality within the process, and an investigation should be initiated to determine the cause.

3. If ß is **equal to 1** then there is a constant mortality distribution rate which suggests that random external events are affecting the failures.

4. If ß is **greater than 1**, the failure rate increases over time which usually represents what is expected from a normal aging process as the component wears out. When ß starts dropping (for instance fouling is increasing) maintenance may be due.

Further analysis can be done by running paretos of the most common failure modes during the commissioning phase or final stage. The biggest issues can be addressed and corrective actions can be put in place to continue to optimize startup and lengthen the time before shutdown.

CHAPTER 13

Implementation

Successful process safety projects have several key elements, which starts by forming a team of subject matter experts. Members should have a strong knowledge of the physics and chemistry of the processes.

Team: They know the potential data sources, and how to access the data to be monitored. Teams need to limit the scope of the first phase to a manageable size such as a specific production line or operating unit.

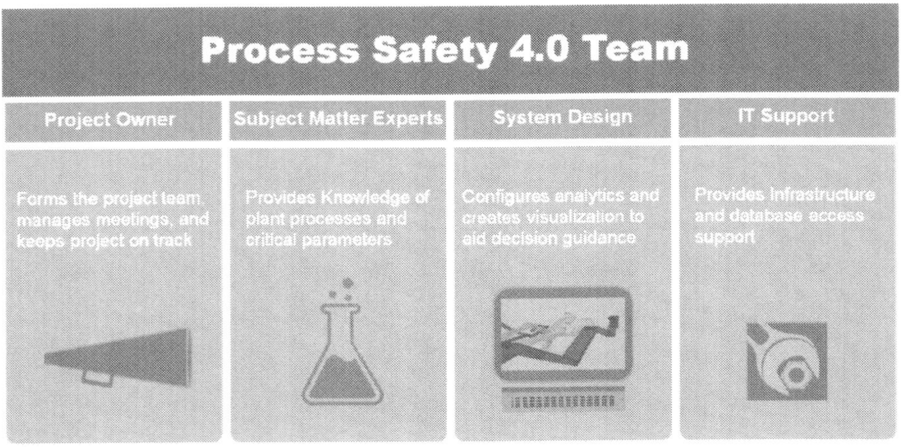

Process Safety 4.0 Team

Project Owner	Subject Matter Experts	System Design	IT Support
Forms the project team, manages meetings, and keeps project on track	Provides Knowledge of plant processes and critical parameters	Configures analytics and creates visualization to aid decision guidance	Provides Infrastructure and database access support

Outputs: Focus on the end first. Looking towards the outputs you are targeting is the best way to stay on track. What are the key decisions which could prevent an incident? How will this knowledge lead to improved process safety? Who will have access to the monitoring system? How will analyzing this data lead to increased knowledge of process safety?

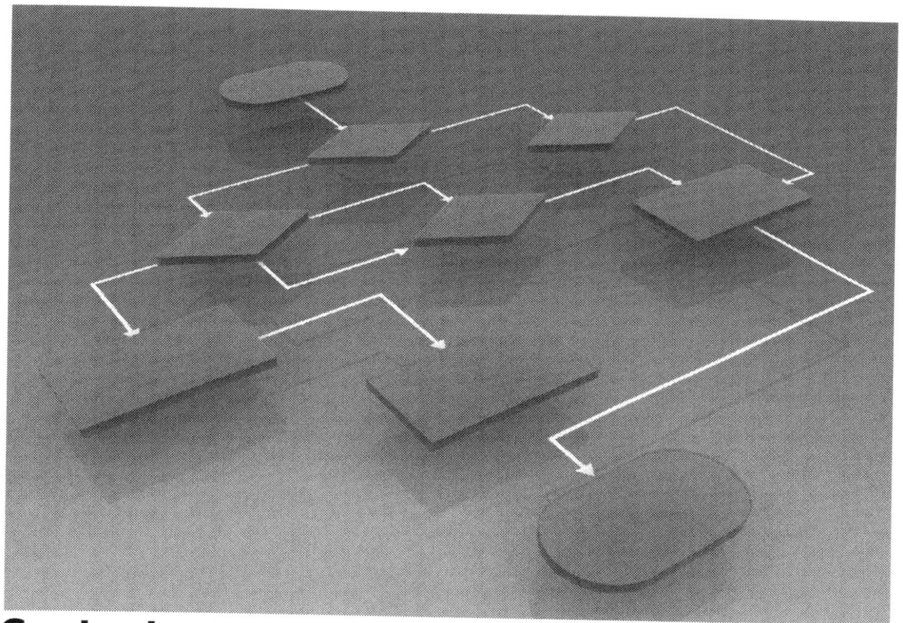

Context: Variables must be organized in context to ensure accurate meaning. This may mean grouping measurements into categories with similar measurement device, operating unit, location, or by a specific metric (such as "pressure drops").

Usually there is a benefit for creating two time horizons to provide different context. One may show hourly data for the last seven days, and another might show data over the last several months.

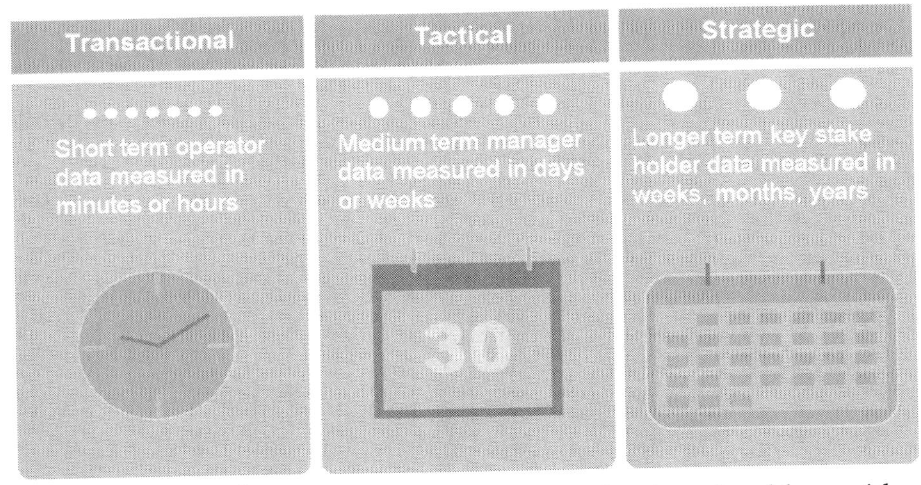

Collaboration: Process safety monitoring should provide a shared view among several key stake holders. Tracking preventative actions, corrective actions and assignable causes is important for shift change communication, supervisor review, and longer term analysis for continuous improvement.

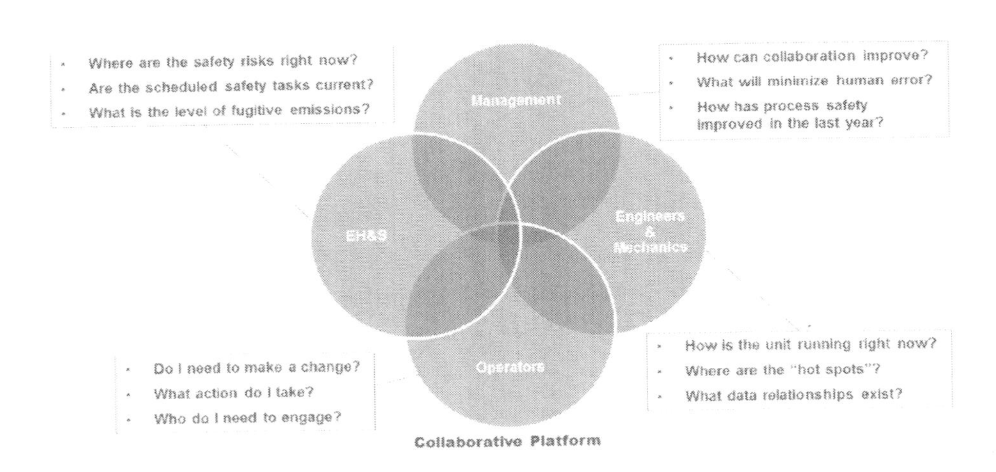

CHAPTER 14

Developing a DHRO Culture

The technology and expertise exists today to build Process Safety 4.0, but to ensure success organizations must build on their foundational culture. Companies need to focus on developing culture around becoming a DHRO. Otherwise they will struggle to find the best areas where process safety can be improved and costs saved.

Most operators believe that their job is to optimize production, and process safety has already been handled by others. Plant managers know that thousands of process alarms are going off daily which are just getting acknowledged or ignored. They worry about which combination and how many of those non-critical alarms will make them vulnerable for disaster. How many at corporate headquarters are helping to drive the "mindfulness" of a DHRO culture?

Typical day in a 4.0 plant:

8:00 Plant manager and plant's data engineer review with shift leader the previous night's events and make decisions on reactive and optimization actions.

9:00 The Industry 4.0 system engineer of a central department meets with plant personnel to make them aware of two new optimization algorithms now available for use.

10:30 Section "x" of the plant is taken out of service for two hours to repair a pump with signals that indicated a bearing problem which would have failed in 3 to 5 days (over the weekend).

10:49 An alarm indicates sensor comparisons with inline measurements are inconsistent with the plausibility algorithm. The alarm provides the operator with the potential causes and best action to follow. The "run-plant" engineer who was also notified checks in with the operator to discuss the decision.

A process engineer may work at a company for 20 years using Excel and other tools to analyze trends and gain manufacturing insights. When they leave the company, IT wipes their hard drive clean and gives the computer to the next person. Any accumulation of process knowledge walks out the door with the previous engineer. An Industry 4.0 platform should accumulate process safety learnings and build objective organizational knowledge along with individual knowledge.

Operators are frequently faced with many difficult behaviors: multiple alarms going off at once, time pressure, walking the line, etc. An Industry 4.0 platform can change the culture of the plant by enabling correct behaviors with better process understanding and decision guidance. The system will provide early warnings by identifying leading indicators, specific direction, and safer actions. New risks, which may not have been recognized in the past because of cognitive bias, are now identified through impartial analytics.

This requires the whole organization to become more consciously competent. Here are some of the challenges to recognize when building a DHRO culture:

Normalization of Risk: Whatever level of risk that exists on a site becomes normal over time. In locations where adverse conditions are the norm people frequently fail to report issues. Risk levels increase even when incidents do not occur as people no longer pay attention to many threats. This becomes the norm.

Budgeting Process: A typical budgeting process has each sub-group submit capital improvement plans to the annual budget. This is typically based on the previous year's budget. Plans are reviewed and usually sent back with a mandate to cut a specific percent. Some areas of a plant may defer replacement plans year after year by this process. There is risk that some plants have far more deferred maintenance than others, leading to higher risk for process safety incidents.

Shift Changes: Different shifts may have different ideas about how to run the process. When shift changes occur, settings and valve positions are often changed in what one operator described as a "tug of war" in the aftermath of the Longford Gas Plant explosion in Australia. (Hopkins, Lessons From Longford).

Ensuring strong communication occurs between shift changes is also a major cultural issue. Having a comprehensive system that combines the relevant process safety parameters into a unified view enables an environment for better communication and collaboration.

Audits: What is the purpose of the audits being conducted? Some companies use them to score, certify or rank plants against each other. This does not drive a DHRO culture. Audits should be searching for problems which might result in catastrophic risk.

Certifications may be more of a checklist to see if there is a system for identifying hazards, that they are being documented, and that incidents are being recorded. This does not challenge mindfulness.

A digital system for tracking audit tasks enables a cultural focus on timely follow up. The tasks are assigned fixed timelines and there is group visibility to progress. Key stakeholders are informed by the system when timelines are about to expire and when they have expired.

Anyone who reports a process safety issue should receive communication about the follow up. Major catastrophes have occurred in environments where operators felt management was ignoring or never responding to reported issues. This developed into a culture of less and less issue reporting.

The cultural environment is critical to the success of deploying an Industry 4.0 solution for process safety. Recognizing the potential barriers allows stakeholders to provide appropriate focus.

Conclusion

Process safety is well suited to leverage the design principles of Industry 4.0. The holistic structure addresses the issues found in the Swiss Cheese model and will reduce the risk of major incidents.

With a deliberate "mindful" effort we have the ability to dramatically reduce incidents and disasters in our highly hazardous process industries. By using data to identify early warning signs, plants will become more predictive and keep processes more closely in the optimal parameter window.

Key elements for creating a DHRO culture:

1. Become data oriented
2. Drive risk awareness
3. Focus on accumulating knowledge
4. Catch early warning signals
5. Learn from incidents and other plants
6. Have discipline for not cutting corners
7. Use data driven operational guidance

The culture should include a stronger element of prediction based on data and fix problems before they lead to incidents. This requires a clear mindset for all involved in the organization.

The process safety expert now has a new set of tools at their disposal. The methods used for process safety in the past decades provide a foundation for identifying the critical parameters to monitor. The process safety expert will be able to use data to drive far better decision guidance than ever before.

Terms Used

APC	Advanced Process Control
AIChE	American Institute of Chemical Engineers
BOP	Blowout Preventer
BOW-TIE	A method of risk analysis
C5I	Command, Control, Communications, Computers, Collaboration, and Intelligence
CCPS	Center for Chemical Process Safety
CSB	U.S. Chemical Safety & Hazard Investigation Board
DCS	Distributed Control System
DHRO	Digital High Reliability Organization
EPA	Environmental Protection Agency
HAZOP	Hazard and Operability Study
HRO	High Reliability Organization
IIoP	Industrial Internet of People
IIoT	Industrial Internet of Things
IPL	Independent Protection Layer
KPI	Key Performance Indicator
LIMS	Laboratory Information Management System
LOC	Loss of Containment
LOPA	Layers of Protection Analysis
LPL	Learning Protection Layer
MES	Manufacturing Execution System
MTBF	Mean Time Between Failure
OEE	Overall Equipment Efficiency
OSHA	Occupational Safety and Health Administration

PCA	Principle Component Analysis
PHA	Process Hazard Analysis
PLS	Partial Least Squares
PSI	Process Safety Incident
QRA	Quantitative Risk Assessment
RM	Risk Matrix / Heat Map / Risk Map
SPC	Statistical Process Control
SIL	Safety Integrity Level
SIS	Safety Instrumented Systems
SCM	Swiss Cheese Model

Works Cited

Ben Leubsdorf, David Gauthier-Villars. "Nobel Goes to Economist Richard Thaler." *Wall Street Journal* 10 October 2017: A2.

Broad, William. "High Risk of New Shuttle Disaster Leads NASA to Consider Options." *New York Times* 9 April 1989.

Browne, Janet E. *Charles Darwin: vol. 2 The Power of Place*. London: Jonathan Cape, 2002.

Cox, Louis Anthony (Tony). "What's Wrong with Risk Matrices." 2008.

"CSB BOP Failure Deepwater Horizon." *: http://www.csb.gov/csb-releases-new-computer-animation-of-2010-deepwater-horizon-blowout/.* Chemical Safey & Hazard Investigation Board, n.d.

Darwin, Charles. *The Origin of Species (1859), Philip Appleman edition*. New York: Norton, 1975.

Heinrich, Herbert William. *Industrial Accident Prevention, A Scientific Approach*. McGraw-Hill, 1931.

Hermann, Mario, Tobias Pentek and Boris Otto. *Design Principles for Industrie 4.0 Scenarios*. Koloa, Hi, USA: IEEE, 2016.

Hopkins, Andrew. *Failure to Learn*. 2008.

Hopkins, Andrew. *Lessons From Longford*. 2000.

Huxley, T.H. *Agnosticism: a rejoinder. In Collected Essays vol 5 Science and Christian tradition*. London: Macmillan, 1889.

Iyengar, Rishi. "Searching Questions Asked in the Aftermath of the Tianjin Blasts." *Time* (2015): August 13.

Kaler, James. *The Hundred Greatest Stars*. New York: Copernicus Books, 2002.

Karl Weick, Kathleen Sutcliffe. *Managing the Unexpected: Assuring high performance in an age of complexity*. San Francisco: Jossey-Bass, 2001.

Komorowski, Matthew. "A History of Storage Cost." 2014.

Lauridsen, Kozine, Markert, Amendola, Christou, Fiori. "Assessment of Uncertainties in Risk Analysis of Chemical Establishments." *Riso National Laboratory, Technical University of Denmark (DTU)* (2002): 1-50.

Martin Sedgewick, Angela Wands. "Scottish Power Goal: To make Process Safety risks as visiable as Health and Safety risks." *www.hse.gov.uk/comah/case-studies/case-study-scottish-power.pdf*. www.epsc.org/data/files/pspi_presentations/ScottishPower.pdf: Performance Metrics Outside the Process Industries - Scottish Power, 2012.

Martin, Joel W. "The Samurai Crab." *Terra* 1993, Summer ed.

Penn, Ivan. "Exxon Mobil's outdated equipment and procedures led to Torrance explosion, agency says." *Los Angeles Times* (2017): May 3.

Vaughan, Diane. *The Challenger Launch Decision: Risky Technology, Culture, and Deviance at NASA*. Chicago and London: University of Chicago Press, 1996.

Acknowledgements

The authors would like to acknowledge and thank the very useful contributions of those who helped improve this handbook: Tjaart Molenkamp, Chris O'Byrne, Debbie O'Byrne, Dana Petrusich, Megan Petrusich, Brian Rohrback, Bill Schuller, Josh Shupp, Brock Zauderer, Jody Martin and the Natural History Museum of Los Angeles County.

About the Authors

Jim Petrusich is a Vice President at NWA Software, a company which specializes in real-time analytic monitoring systems in chemical process industries. He has worked on many data aggregation and decision guidance projects with some of the world's largest oil, gas and chemical manufacturers. These deployments have won multiple industry awards including the Manufacturing Intelligence Award (2016, Frost and Sullivan).

Prior to joining NWA in 2010, he was Vice President, Sales and Services for Planar Systems, where he was responsible for large deployments of automated monitoring systems (C4I and C5I) in military, surveillance, transportation, and energy projects with many of the world's largest companies and governments.

Hans Volkmar Schwartz is Vice President, Process Safety at BASF. He had group wide responsibility for process safety at the world's largest chemical company with more than 100,000 employees in over 80 countries until end of 2016, and, after reorganization, is now supporting BASF's Process Safety approach as an executive expert. Schwarz started his career at BASF in 1986 and progressed through the organization with senior roles in Europe and North America, including plant management, project management, technology director and project director with responsibility for supply chain optimization.

His production and technology roles included the development of process safety concepts for some of BASF's most hazardous technologies. Schwarz was given BASF's North American innovation award in 1999.

Schwarz obtained his doctorate in physical chemistry.at Ruprecht-Karls University in Heidelberg, Germany.

INDUSTRY4.0
for
PROCESS SAFETY
HANDBOOK
JIM PETRUSICH
HANS VOLKMAR SCHWARZ

Made in the USA
Columbia, SC
13 December 2017